IMMORTALITY

Nonfiction by Dr. Ben Bova

SPACE TRAVEL

THE CRAFT OF WRITING SCIENCE FICTION THAT SELLS

FIRST CONTACT (EDITOR AND CONTRIBUTOR)

THE BEAUTY OF LIGHT

INTERACTIONS (WITH SHELDON GLASHOW)

WELCOME TO MOONBASE!

STAR PEACE

ASSURED SURVIVAL

VISION OF THE FUTURE: THE ART OF ROBERT McCALL

THE HIGH ROAD

THE SEEDS OF TOMORROW

VIEWPOINT

CLOSEUP: NEW WORLDS (WITH TRUDY E. BELL)

NOTES TO A SCIENCE FICTION WRITER

SCIENCE: WHO NEEDS IT?

THROUGH EYES OF WONDER

WORKSHOPS IN SPACE

THE WEATHER CHANGES MAN

STARFLIGHT AND OTHER IMPROBABILITIES

THE NEW ASTRONOMIES

THE AMAZING LASER

THE FOURTH STATE OF MATTER

PLANETS, LIFE & LGM

IN QUEST OF QUASARS

THE USES OF SPACE

REPTILES SINCE THE WORLD BEGAN

GIANTS OF THE ANIMAL WORLD

THE MILKY WAY GALAXY

IMMORTALITY
HOW SCIENCE IS EXTENDING YOUR LIFE SPAN— AND CHANGING THE WORLD

DR. BEN BOVA

AVON BOOKS ◆ NEW YORK

AVON BOOKS, INC.
1350 Avenue of the Americas
New York, New York 10019

Copyright © 1998 by Ben Bova
Cover design and illustration by Amy Halperin
Inside back cover author photograph by Eric Strachan
Interior design by Kellan Peck
Published by arrangement with the author
ISBN: 0-380-79318-0
www.avonbooks.com

The Avon Books hardcover edition contained the following Library of Congress Cataloging in Publication Data:

Bova, Ben, 1932–
    Immortality / Ben Bova.—1st ed.
      p.   cm.
    Includes bibliographical references and index.
    1. Longevity—Popular works. I. Title
RA776.75.B68   1998   98-3950
612.6'8—dc21   CIP

First Avon Books Trade Paperback Printing: January 2000
First Avon Books Hardcover Printing: September 1998

AVON TRADEMARK REG. U.S. PAT. OFF. AND IN OTHER COUNTRIES, MARCA REGISTRADA, HECHO EN U.S.A.

Printed in the U.S.A.

OPM 10 9 8 7 6 5 4 3 2 1

To the memory of Isaac Asimov,
beloved friend and treasured mentor,
with more thanks than can ever be expressed.

Acknowledgments

This book could not have been written without the generous assistance of Dr. Lawrence E. Allred, Philip Brennan, Dr. Martha Davila-Rose, Robert C. W. Ettinger, Dr. Kenneth Jon Rose, Scott Spooner, Bruce Sterling, and many others. They have all supplied valuable information and insights and they have my enduring thanks. They are not responsible, however, for any errors of fact or judgment displayed herein. The factual accuracy of this book, and the conclusions reached, are no one's responsibility but my own.

Nothing is too wonderful to be true.
　　　　　　　　—Michael Faraday

Contents

Part One
The Scientific Evidence

Part Two
The Impact of Immortality

Preface: A Prophet Is Not Without Honor . . .

We may learn to deprive large masses of their gravity, and give them absolute levity, for the sake of easy transport. Agriculture may diminish its labor and double its produce; all diseases may by sure means be prevented or cured, not excepting even that of old age [emphasis added], and our lives lengthened at pleasure even beyond the antediluvian standard. . . .

—BENJAMIN FRANKLIN

∞ BEN FRANKLIN'S VISION OF THE FUTURE, WRITTEN IN A letter to the scientist Joseph Priestly in 1780, is a striking example of optimistic prophecy based on the hopes of scientific advances.

Similarly, this book offers an optimistic prophecy based on the promise of scientific advances: Physical immortality is within sight. There are people living today who may extend their life spans indefinitely.

Prophecy can be a tricky business, as many weather forecasters and stock market analysts can attest. Moreover, human nature tends to accept gloomy prophecies as probably correct, or nearly so, while optimistic prophecies are usually harder to accept and are often greeted with: "That's too good to be true!"

How well did Ben Franklin's prophecy turn out?

Franklin's expectation of an antigravity device that can "deprive large masses of their gravity, and give them absolute levity" has not been realized, yet modern trucks, trains, forklifts, airplanes, and other cargo carriers haul large masses across the globe without the need for the backbreaking labor by men or animals that existed in Franklin's day.

If Franklin was overly optimistic in regard to cargo hauling, he was not hopeful enough about agriculture. Using farm machinery and chemical fertilizers and pesticides, modern agriculture produces more than a hundred times more food per acre than farmers could in the eighteenth century with a fraction of the human labor.

In the field of medicine, many of the diseases that struck terror in Franklin's time—from smallpox to pneumonia—have either been eradicated entirely or brought under control by antibiotics. Thanks to the development of antiseptics, anesthetics, and our growing knowledge of the human body, surgery has evolved from bloody limb hacking to successful organ transplants.

As Franklin correctly foresaw, these improvements in medicine and surgery have led to lengthened life spans. No one has approached the "antediluvian standard" of Noah's supposed 950 years or Methuselah's 969, but people in the industrialized world today live far longer, far healthier lives than was possible in Franklin's era.

Today we are approaching the point where human beings can and will live for hundreds of years or even more. Physical immortality, if not yet within our grasp, is certainly within our vision.

As Carl Sagan often said, "Extraordinary claims require extraordinary evidence." The pages of this book will offer evidence from research laboratories around the world that human immortality is no longer a fantasy or a dream but may be achievable within this generation.

First, however, perhaps a few words about my own record as a prophet may be in order. I do not seek to praise myself but to show that in the field of scientific prognostication I am not a complete neophyte.

I have been writing fiction and nonfiction about the future since 1949. My nonfiction books deal mainly with astronomy and physics, although my interest in the possibilities of extraterrestrial life has brought me into several areas of biology and genetics, particularly molecular biology. I am not a scientist by training; my degrees are in journalism, communications, and education.

Most of my novels and story collections are categorized as science fiction, yet a good many of the earlier ones can now be read as history. My very first novel, written in 1949–50, was not published because editors at that time thought the plot was too fanciful. The novel was based on the idea that the Soviet Russians go into space before the United States, so the United States sets up a crash program to land Americans on the Moon before the Russians can get there. The space race of the 1960s vindicated my vision, but by then my novel was dated.

Among other predictions I have made are:

- the invention of solar power satellites (in a 1960 short story)
- the discovery of organic chemicals in interstellar space (1962)
- virtual reality (1962)
- the Strategic Defense Initiative, a.k.a. "Star Wars" (1976)
- international peacekeeping forces (1988)
- electronic book publishing (1989)
- the discovery of life on Mars (1992)
- the discovery of water ice at the south pole of the Moon (1996)

As baseball's Dizzy Dean said, "It ain't bragging if you can do it." I present this list as evidence of my credentials to make a prediction that will seem far-fetched, even outlandish to most readers—and most scientists:

Human immortality is within our grasp.

PART ONE

THE SCIENTIFIC EVIDENCE

∞

It is the great glory and also the great threat of science that anything which is possible in principle—which does not flout a bedrock law of physics—can be done if the intention to do it is sufficiently resolute and long-sustained.

—SIR PETER MEDAWAR

I

Extending the Human Life Span

The days of our years are threescore years and ten . . .
— THE BOOK OF PSALMS, PSALM 90:10

∞ THE FIRST IMMORTAL HUMAN BEINGS ARE PROBABLY LIV-
ing among us today. You might be one of them.

There are men and women alive today who may
well be able to live for centuries, perhaps even extend
their life spans indefinitely. For them, death will not
be inevitable.

As the American immunologist William R. Clark put
it, "Death is not inextricably intertwined with the definition of life." Just because human beings have always died
does not mean that they always will die.

Accidents and violence will not disappear, of course.
People will still be vulnerable to poor judgment, bad luck,
and evildoers. But death from old age, death as the inescapable end of life, will become a thing of the past, a dark
memory of primitive days.

You might be one of the immortals. Particularly if you
are less than 50 years old, in reasonably good health, and
live a moderate lifestyle, you may live for centuries or
longer. If you smoke, or drink to excess, or take narcotics,

or are involved in hazardous work or play, your chances of extending your life span are of course reduced.

But if you have a normal life expectancy today, the medical and biological advances that will be achieved over the next ten to twenty years will probably allow you to live long past 100; and the longer you live, the more knowledge that our biomedical scientists glean, the farther and farther your life span will be extended.

Time is on your side.

I am not talking about living as a feeble, sickly old person. You may well be able to maintain the youth and vigor of a 50-year-old indefinitely or perhaps even reverse the effects of aging and become physically younger.

Few reasonable scientists would agree with this prediction. They know too much about the difficulties of their work, the intricacies of the human body, the vast seas of unknowns that stretch ahead of them to be so glib as to accept the idea of imminent human immortality and the reversing of the aging process. But scientists are usually not the best predictors of their own futures. For example:

In 1903 the American astronomer Simon Newcomb wrote, "Aerial flight is one of the great class of problems with which man can never cope. . . ." Two months later the Wright brothers flew at Kitty Hawk.

In 1933 the British Nobel prize-winning physicist Ernest Rutherford, soon after the first splitting of the atomic nucleus, predicted, "The energy produced by the atom is a very poor kind of thing. Anyone who expects a source of power from the transformation of these atoms is talking moonshine."

Albert Einstein agreed: "There is not the slightest indication that [nuclear] energy will ever be obtainable."

Hiroshima was only a dozen years away.

Nor are politicians and media pundits any better at prognostication.

In 1844 Senator Daniel Webster railed against the acquisition of California by the United States: "What do we want of the vast worthless area? This region of savages and wild beasts, of deserts, of shifting sands and whirl-winds of dust, cactus and prairie dogs? . . . What use can we have for such a country?"

On January 18, 1920, *The New York Times* chided rocket pioneer Robert Goddard and proclaimed its editorial opinion that rockets could not work in space because "of the need to have something better than a vacuum against which to react."

5

Perhaps our best guide in examining the future is the maxim of writer Arthur C. Clarke: "When a distinguished but elderly scientist states that something is possible, he is almost certainly right. When he states that something is impossible, he is very probably wrong."

Today hardly anyone in science, medicine, or govern-ment has faced up to the inevitable challenges—and op-portunities—of vastly extending human life spans. In the United States, political arguments rage over the financial health of the Social Security system and Medicare, the government's medical insurance program for persons over 65 years of age. What will happen when the "elderly" live far beyond the century mark and are healthy and vigorous enough to keep on going, draining every pen-sion fund in the nation?

Possibly, healthy and vigorous "golden agers" will es-chew retirement in favor of new careers, but as the laws now stand—not only in the United States but in every industrialized nation of the world—most people can retire at age 65 (or younger) and collect their pensions and med-ical benefits as long as they live.

Clearly, the biomedical breakthroughs that will come over the next few decades are going to shatter these social arrangements and force us to deal with the facts of greatly extended life spans and virtual immortality.

Life Expectancy and Life Span

There are more than fifty thousand centenarians in the United States today. Americans live longer after the age of 80 than any other nation's population. And the number of Americans who live past 100 is doubling every ten years. People over 85 are the fastest-growing segment of the nation's population.

If these trends continue, by the time today's 30-year-olds reach 80, there will be some 1.6 million centenarians in the United States. And today's trends will not merely continue, they will accelerate, thanks to the advances in biomedical knowledge coming from research laboratories all over the world.

Psalm 90, a prayer that the Bible attributes to Moses, says:

*The days of our years are threescore years and ten; and
if by reason of strength they be fourscore years, yet is
their strength labor and sorrow; for it is soon cut off,
and we fly away.*

Thus the concept that a human being's "natural" life span is at best somewhere between 70 and 80 years has been with us for a long time.

Yet Moses himself, according to the Bible, lived to be 120. Modern scholars do not accept that figure, but vastly elongated life spans dot the Old Testament. Methuselah's age is given as 969 years; Noah's, 950. The chances are that those numbers actually refer to months, not years. That would make Methuselah almost 81, Noah slightly more than 79. Still, in an age when most people did not live to see 30, those were remarkable life spans.

In Homer's *Iliad* the oldest of the Greek kings fighting at Troy is white-bearded Nestor. He is so aged that haughty Agamemnon and the other kings often turn to him for advice. Yet from the evidence within the saga, Nestor cannot be more than 50. That was a hoary old age in the days when warriors such as Achilles were cut down in their teens.

6

As recently as the year 1900, life expectancy at birth for the average American male was only 48.3 years; for the average American female, 51.1.

And life expectancy in the United States was about as good as anywhere on Earth. Australia was actually the world leader in 1900, with life expectancies at birth for males at 53.2 years and females at 56.8. In Europe the best life expectancy figures came from the Scandinavian nations of Sweden (52.8 and 55.3), Norway (52.3 and 55.8), and Denmark (51.6 and 54.8). Japan was only 42.8 and 44.3; Spain 33.9 and 35.7.

The biggest reasons for this low life expectancy (low by today's standards) were childhood diseases that killed infants in their early years and unsanitary environmental conditions that gave rise to infectious diseases such as cholera and tuberculosis.

By the middle of this century, 1950, life expectancy at birth for Americans had increased to 66.0 for males and 71.7 for females, a gain of eighteen years for men and twenty for women. Similar gains were made in the other industrialized nations.

In 1996 life expectancy at birth had risen to 75.7 years for the average American male and 82.7 for the average American female. Moreover, today average life expectancy actually improves with age, as Table 1 shows.

Table I
Life Expectancies

If you make it to age:	You can expect to live another:			
	WHITE MALES	WHITE FEMALES	NONWHITE MALES	NONWHITE FEMALES
50	27.0 yrs.	31.7 yrs.	24.3 yrs.	20.5 yrs.
70	12.3 yrs.	15.3 yrs.	11.5 yrs.	14.3 yrs.
85	5.2 yrs.	6.4 yrs.	5.1 yrs.	6.3 yrs.

Note: These figures are for the United States only.

What this means is that the longer you live, the longer you can expect to live. Up to a point. At birth the average American male can expect to live 75.7 years. But if he makes it age 50, he can expect to live to 77. If he lives to be 70, chances are he will reach 82 and beyond; if he reaches 85 (the highest age for which the National Center for Health Statistics keeps records), he can expect to see 90.

Women live somewhat longer—and always have, wherever actuarial statistics have been kept. Partly this is because males are more prone to violent death, especially in their younger years. Men also are more likely to have jobs that are more stressful or hazardous, although as women continue to move into the workplace, this effect is beginning to even out. It is especially noteworthy that death from lung cancer has risen among American women to nearly equal the rate of male fatalities from lung cancer, thanks mainly to the increase in women who smoke cigarettes. This is one of the unwelcome aspects of the feminist movement.

Throughout the twentieth century, however, average life expectancy has almost doubled.

There are many reasons behind these trends, but the chief factors are better nutrition and cleaner environmental conditions. Despite the strident wails of those who claim that global environmental pollution is killing us, humans in all the industrialized nations (where accurate records are maintained) are living longer.

According to Kevin Kinsella of the Center for International Research, U.S. Bureau of the Census:

Expansion of public health services and facilities combined with disease eradication programs to greatly reduce death rates, particular among infants and children.

Yet while life *expectancy* has nearly doubled in this century, the maximum human life *span* still seems to reach little farther than "threescore years and ten . . . [or] four-score years." More men and women are reaching their seventies and eighties, but relatively few go much beyond that age.

There have been claims, from time to time, of individuals who have lived much longer.

Englishman Thomas Parr lies buried in Westminster Abbey on his unsubstantiated claim that he was more than a century and a half old. He died in 1635, allegedly at the age of 152. Subsequent investigation showed that he was a fraud, but the wealthy society of London apparently wanted to believe him, and he was as famous among them as Madonna or Michael Jackson is in our society today.

In the 1960s tales filtered out of the Soviet Union of hardy farmers in the Caucasus Mountains region who claimed to be 150 or older. Yogurt and clean air were purported to be the key to a long life, a factor that led major yogurt manufacturers to accept their claims uncritically. But there were no reliable records of their births, and at least one such old-timer was exposed as a fraud—and a World War I army deserter—who was actually "only" 78.

In the 1959 Soviet census, 97 percent of all the centenarians in the U.S.S.R. came from Georgia, which had only 2 percent of the Soviet Union's population. But Joseph Stalin was a Georgian, and apparently he liked to believe that his people lived longer than anyone else. The Soviet apparatchiks were only too happy to confirm their leader's belief. Stalin himself died at the age of 74.

The oldest human being whose age is reliably recorded was Jeanne Louise Calment of Arles, France, who was born on February 21, 1875, fourteen years before the Eiffel Tower was built, and died on August 4, 1997, at the age of 122 years, 5 months, and 14 days. She remembered selling pencils to a struggling young artist named Vincent van Gogh.

Is there a natural limit on human life span? Is there something built into our bodies, into our genes, that prevents us from living indefinitely?

If there is, can modern science find ways to break through this natural limit?

The answers seem to be: Yes, yes—and yes.

10

Birth Rate vs. Death Rate

The dramatic improvement in life expectancy over the past century has come mainly from lowering the death rate among children. There has been no such lowering of the death rate among the elderly—so far.

In 1993, for example, in the United States, 7.7 white male infants under the age of 1 died per thousand 1-year-olds. The death rate for nonwhite males under 1 was more than double that figure: 15.9. Yet as recently as 1960, the death rate for white American males under 1 was 26.9 and a staggering 51.9 for nonwhites. In all cases, females fared slightly better.

Clearly, infant mortality has been reduced a great deal. But what about mortality among the elderly?

The death rate doubles, roughly, every ten years of life from age 20 to 80. Among white American males 85 and over, 184.4 per thousand died in 1993. In 1960 the number was 217.5 per thousand. The ratios are similar for white women and nonwhites of both sexes.

Although many, many more men and women are living into their seventies and eighties, relatively few live beyond. Better nutrition, better sanitation, and better medicine have helped the young far more than the elderly.

Until now. In recent years geriatrics has become a

"hot" specialty in the medical and sociological professions. Physicians, psychologists, and social workers are more and more concerning themselves with the physical, medical, psychological, economic, and social problems of the elderly. Although geriatric care is mainly concerned with helping the elderly to live as comfortably as possible, research scientists are examining the physical and genetic roots of old age. Why do we become progressively more feeble and more prone to certain diseases as we grow older?

And research scientists around the world are turning their attention to a fundamental question:

Is there truly a built-in, biological limit on how long a human being can live?

With all our wonder drugs and purified water and nutritious food, have we hit a "wall" at the far end of the human life span? Are we in a situation where, no matter what we do, most people will die after their threescore and ten years, and no one will make it much past 100?

The answers appear to be: Yes, maybe—and *no*. Even if there is a "wall" that limits the human life span, scientists are learning how to break through it.

How long can we live, then?

In an expression borrowed from the world of aviation, how far can we push the envelope?

The Four Eras of Medicine

Since the dawn of human history, people have sought ways to reduce or eliminate the pain and debilitation of disease and ways to extend their life spans. The following sidebar gives the humorous view of an Internet cynic on the subject.

A SHORT HISTORY OF MEDICINE

"Doctor, I have an earache."

2000 B.C.: "Here, eat this root."

1000 B.C.: "That root is heathen. Say this prayer."

1850 A.D.: "That prayer is superstition. Drink this potion."

1940 A.D.: "That potion is snake oil. Swallow this pill."

1985 A.D.: "That pill is ineffective. Take this antibiotic."

2000 A.D.: "That antibiotic is artificial. Here, eat this root!"

Humor and cynicism aside, the quest for health and long life has been part of the human drama since our ancestors first realized the finality of death.

The Epic of Gilgamesh dates back to the dawn of civilization. It is one of the earliest tales we know of, predating the Biblical story of the Flood. In it, the hero-king Gilgamesh seeks the secret of immortality. Unsuccessfully.

By the time that writing was invented, some five or six thousand years ago, people had learned that certain natural substances—usually from plants—could alleviate pain or even cure disease. Certain roots or berries could ease stomach cramps, for example. Chewing on the bark or leaves of a certain type of willow tree relieved pain.

Over many hundreds of generations, preliterate peoples accumulated considerable pharmacological knowledge, the kind that we would now consider *folk medicine*. No one knew why these remedies worked, but they knew that they did work—some of the time, at least.

Ignorance is almost always accompanied by fear, and humans have a tendency to try to explain the unknown, to drive away that fear. Thus tribal shamans attributed the efficacy of their "home remedies" to supernatural causes. Spirits or gods created the cure, through the shaman's knowledge not only of medicinal herbs but also through the proper rituals that had to accompany the treatment.

Today, while we may smile at such primitive superstition, physicians know that the placebo effect is very real: In some cases, the patient gets better because he or she believes the "medicine" administered by the doctor actually works, even when the "medicine" may be nothing more than sugared water, a harmless, impotent placebo.

Slowly, over the centuries, knowledge of disease and the human body improved. Yet medicine was still more or less of a cookbook affair: "Try this because it usually works."

It was not until the nineteenth century that medicine entered its second era. Organized science, particularly chemistry, began to examine *why* certain substances produce certain results in the human body. For example, biochemists learned that acetylsalicylic acid relieved pain and inflammation. You didn't have to chew the bark of a tree; you could refine the acetylsalicylic acid in a chemistry lab and press it into convenient tablets: aspirin.[1]

Vaccines were developed to build up an individual's immunity against certain diseases. In 1796 Edward Jenner began to inoculate patients in England with cowpox serum to protect against deadly smallpox. Seventy-five years earlier, Cotton Mather took a much more dangerous path by inoculating his own son with pus from a smallpox victim (see Witchcraft and Science, page 14).

Although no one then understood how vaccination worked, it worked, and once cowpox inoculations became

13

---

[1]Aspirin was first produced by German chemist Felix Hoffmann in 1890. Working for the German firm of Bayer & Co., he refined salicylic acid and acetyl from coal tar. Since salicylic acid comes naturally from the plant meadowsweet (scientific name: *Spirea ulmaria*), Hoffmann coined an acronym out of *a* for acetyl, *spir* from *spirea*, and *in* for reasons known only to himself. The result: aspirin.

WITCHCRAFT AND SCIENCE

Cotton Mather, the famous Massachusetts preacher and writer, was nearly killed by a bomb in 1721 because he supported the idea of inoculation against smallpox.

Son of Increase Mather, who greatly influenced the Salem witchcraft trials a generation earlier, Cotton Mather believed that witches might exist. Yet he also believed in science, and when his own son came down with smallpox, he inoculated the boy with pus from a smallpox victim. He had heard of the African practice of inoculation from one of his slaves and had read about its use in Turkey. The son, also named Increase, nearly died but ultimately survived.

Many in the Boston area were outraged. Mather's fellow ministers proclaimed that smallpox was a punishment from on high and any attempt to circumvent it was interfering with divine will. There were riots and threats of hanging. A crude bomb was hurled through a window of Mather's house; fortunately, it did not explode.

The *New England Courant* ran a series of satirical essays against Mather and inoculation, written by "Silence Dogood," pen name for Benjamin Franklin. At age 16, it was Franklin's first published writing.

Franklin grew wiser with age and became a respected scientist—among his other accomplishments.

widespread, the smallpox plagues that ravaged human societies became a thing of the past. However, since knowledge of the human immune system was at best primitive, not everyone accepted the idea that inoculation could prevent disease. Nearly two centuries after Jenner, playwright George Bernard Shaw still scoffed at the idea

of vaccination. Today arguments rage in the medical pro-
fession over the practice of homeopathy, which employs
inoculations of vanishingly small amounts of substances
in attempts to build immunity to diseases.

Meanwhile, biologists were using their microscopes
(as important to their work as telescopes are to astrono-
mers) to find that the human body is made up of tiny
cells, 100 trillion of them per person, on average: skin
cells, muscle cells, nerve cells, specialized cells for each
part of body.

Following pioneers such as the French chemist Louis
Pasteur, scientists began to unravel the underlying *causes*
of infectious diseases: microscopic organisms that attack
the human body's cells. The so-called "germ theory of
disease" provided, for the first time, an understanding of
why we get sick. Microbes invade our bodies and attack
our cells.

The "germs" that nineteenth-century biologists discovered were single-celled organisms. Later even smaller disease agents were found: viruses.

Biologists and chemists started to seek out specific
chemical substances that could destroy invading microbes.
Early in the twentieth century, the German biochemist
Paul Ehrlich coined the term *chemotherapy*, meaning the
use of chemicals against infectious disease. Ehrlich sought
and found, in 1910, a "magic bullet" that could cure syphilis: a specific chemical compound that destroyed the invading syphilis microbes without harming the patient.

By 1932, the first drugs effective against the microbes
that led to blood poisoning were developed, and by the
end of World War II, penicillin (actually discovered in
1928) became the first of a host of antibiotics that are the
mainstay of the modern physician's pharmaceutical arse-
nal against disease and infection.

Organized research and health systems on a national
scale eliminated or greatly reduced many of the infectious

diseases that had plagued human society since time immemorial: pneumonia, influenza, smallpox, bubonic
plague, tuberculosis, polio, diphtheria, whooping cough—
these and many others fell before the advance of biomedical knowledge and improved health care.

Meanwhile, all through the nineteenth and twentieth
centuries, the less glamorous but equally important development of sanitary engineering made tremendous contributions around the world to eliminating diseases such as
cholera brought on by impure drinking water and other
environmental factors. Draining swamps where mosqui-
toes bred removed the scourges of malaria and yellow
fever from cities such as Rome and Washington, D.C.

These improvements in sanitation and pharmacology
were behind the dramatic increase in average life expectancy that marked the second era of medicine.

In 1953 medicine entered its third era, delving deeper
into the heart of the cell, into the realm of molecules (see
Molecules, page 17). James Watson and Francis Crick,
American and British molecular biologists respectively,
discovered the double-helix structure of *DNA*—the giant
molecule that directs the genetic program for every cell
of every living organism.

Deoxyribonucleic acid (DNA's full name) is the molecule of genetics. Genes are parts of the giant DNA double-
helix molecule, like pearls strung on a string. DNA contains
the blueprint for each individual organism and directs the
workings of each cell in the organism. Your every physical
attribute, from the color of your eyes to the shape of your
toes to the number of neurons in your brain, is governed
by the DNA in your cells. The attributes you pass on to
your offspring, your genetic heritage, is determined by
your individual DNA and that of your mate.

Many diseases that assail us are *genetic* in nature: They
happen only to individuals whose DNA has been altered
from normal. Cancer, diabetes, cystic fibrosis, sickle-cell

MOLECULES

The cells of our bodies, as well as free-living bacteria and other single-celled organisms, exist in a microscopic world and deal with individual molecules the way we deal with shopping bags full of groceries—or garbage.

Molecules are groups of atoms bound together. The oxygen you breathe, for example, is in molecular form, two oxygen atoms bound together: O_2 in the chemist's shorthand. The nitrogen that makes up nearly 80 percent of our air is also bound into two-atom molecules: N_2. Common table salt is a two-atom molecule of sodium and chlorine: NaCl.

Water—H_2O—is a three-atom molecule: two hydrogen atoms and one oxygen. So is the carbon dioxide that you exhale: CO_2. *Organic* molecules (that is, molecules that contain long chains of carbon atoms linked together) can be much larger. The gasoline that powers your automobile consists of long chains of carbon-based molecules containing hundreds of atoms.

The proteins that make up your body are organic molecules that contain thousands of atoms. Proteins are usually compressed and folded like twisted Slinky toys, which helps them to move without hindrance through the pores and pumps of individual cells.

The largest molecule of them all is DNA, which is composed of millions of atoms.

anemia, and many other afflictions are caused either partially or entirely by defects in the genes.

Aging itself may be a genetic "disease."

With their growing understanding of the molecular basis of DNA, biomedical scientists have begun trials of gene therapy, attempts to alleviate or even eliminate genetic defects and diseases by replacing faulty genes with healthy ones. This work, in its infancy today, is the hallmark of medicine's third era.

While researchers seek to understand the workings of DNA, other scientists have begun to probe the very nature of death itself. Biologists have uncovered intriguing—perhaps frightening—evidence that our DNA includes "death genes," sets of instructions that command individual cells to die. What physicians call *senescence* and we call old age may be specifically ordered by a particular set of genes in our DNA.

The fourth era of medicine is just beginning. With the knowledge so painstakingly gained of how the cells of our bodies work and how the DNA-based genes within those cells operate, *biogerontologists* (the scientists who study human aging) are starting to examine the question of life extension. They are seeking out the workings of programmed cell death and attempting to discover whether or not there is a built-in "wall" that limits the human life span, and if there is, how that "wall" may be circumvented.

Whether they realize it or not, today's researchers are reaching toward human immortality: the end of death as the unavoidable end of life.

Most scientists would be shocked to think that the work they are doing in their laboratories will inevitably lead to the vast extension of human life spans, to physical immortality. They are concentrating on individual investigations, focusing on specific research programs.

Yet that is where today's research is leading us: to the extension of human life spans well past the century mark. And, of course, the more scientists learn about how our cells and our genes work, the farther they will push the limits of the human life span. Immortality is the goal, whether admitted openly or not.

That is what this book is about.

This book is partly a survey of what is known today and partly an informed speculation about what may be possible tomorrow. I do not intend to present a painstaking review of current biomedical research; the books and articles listed in the reference section do that in far more detail than I could. Instead, I want to show the broad outlines of what is known today and what is under investigation at leading research laboratories around the world. From that information, I will extrapolate into the future to demonstrate why I have come to the conclusion that human immortality is within sight and offer some thoughts on the new society that will result from this startling change.

I firmly believe that today's research will lead—ultimately, inevitably—to human immortality.

And that will bring about the greatest challenges—and opportunities—that humankind has ever faced.

2

How We Age

*We owe God a death . . . he that gives it this year is quits
for the next.*

—WILLIAM SHAKESPEARE,
King Henry IV, Part II

∞ FOR A SUBJECT OF SUCH VITAL INTEREST TO EACH OF US,
surprisingly little is known about aging. Until the past
few decades, hardly any scientific studies of aging had
been done.

Yes, there was research into the causes and possible
alleviations of various diseases associated with old age,
diseases such as cancer, atherosclerosis, Parkinson's, and
so on. And all through history, men sought ways to maintain their sexual powers into old age. The search for the
Fountain of Youth has been predominantly a male search;
women face the inevitability of menopause and the inescapable fact that their childbearing years will end with
rather different attitudes from aging men who want to
continue to be "sweetly dangerous" with the opposite sex.

But most physicians and philosophers accepted the
idea that Shakespeare epitomized in *King Henry IV, Part
II*: Death is inevitable and old age is its harbinger.

"Old age" was often given as the cause of death in
earlier generations, even when "old" meant no more than

60! The prevailing thought was that people "wore out" as they went through life; the harder and "faster" they lived, the sooner they wore out and died.

On the other hand, there were those who warned against "rusting out," becoming so inactive that the body's machinery just ran down and stopped altogether.

Again, the hidden agenda here was mainly men who were particularly concerned about the lowering of their sexual prowess as they aged. They understood that exercise helped to prolong vigor and formed the concept of "Use it or lose it." In the end, however, they lost it anyway if they lived long enough.

Most didn't.

Do We Wear Out?

Aesop's fable of the tortoise and the hare can be read as a parable about aging.

Scurrying little critters such as mice live for four years at most. Rabbits and hares rarely make it through two decades. Lumbering Galápagos tortoises can keep plodding along for up to two centuries. Would mice or rabbits live longer if they would take it easy and slow down to a turtle's pace?

They simply can't. Mammals, which maintain a steady internal temperature, must burn energy faster than "cold-blooded" reptiles such as tortoises. Small mammals burn energy faster and have higher body temperatures than large mammals. Their *metabolic rate*—the rate at which they burn energy—is intrinsically high. They have to eat a lot just to keep those inner fires going. Tiny shrews, for example, must consume roughly their own body weight in food every day. "They can starve to death during a long nap," as zoologist and aging specialist Steven N. Austad puts it. Turtles and other reptiles can go for weeks without eating.

Yet many types of birds have quite high metabolic rates but live as long as humans and even longer. Parrots have life spans exceeding our own. Gulls and petrels that have been watched by biologists for half a century or more seem hardly affected by age at all; they still mate and nest, decade after decade, while the scientists studying them grow old and gray.

Despite their extremely high metabolic rates, many species of birds can live very long lives. So metabolic rate alone is not the key to aging. We do not inevitably wear out.[2]

So what causes aging? What happens inside our bodies that makes us age?

Diseases do not age us. Diseases can be thought of as attacks by outsiders who invade our bodies or as malfunctions of the body's own mechanisms. If an automobile suffers a flat tire or a dent from a collision, that is not necessarily a sign of aging. But an auto can succumb to rust over the years.

Aging itself is mainly due to two processes that are going on within your body every moment of every day of your life. One of them is very much akin to rusting.

Free Radicals

Oxygen.

It is the stuff of life. Without oxygen to breathe, we die within minutes. Oxygen feeds our inner fires. Without

---

[2]Among the longest-lived animals are the Pacific rockfish, which lives to 120, and the quahog clam, which can go to 200 years. Plants have much longer life spans. The saguaro cactus of Arizona lives for centuries, and the scrubby creosote bush can survive for thousands of years. In 1997 Tasmanian botanists discovered a species of King's holly that they estimate to be 43,000 years old, dethroning the previous record holder, a 13,000-year-old box huckleberry in Pennsylvania.

it, the 100 trillion cells of our bodies could not convert the food we eat into energy.

Yet nothing in the universe comes without a price. Oxygen is a corrosive element, quite capable of yellowing the pages of a book until they crumble or turning a sturdy bar of iron into a pile of rust. If you breathe pure oxygen at normal room pressure, you will suffer chest pain, coughing, and a sore throat within six hours. This is one of the reasons why astronauts are supplied with oxygen at reduced pressure.

Hospitals found that premature babies placed in incubators that were filled with oxygen-enriched air went blind because of oxygen damage to their retinas.

Lungs and eyes are not the only parts of the body that are prone to oxygen damage. Our cells are constantly being damaged by molecules called *free radicals* (see Free Radicals, p. 24). We are literally rusting away.

The most damaging free radicals are built around oxygen atoms. Oxygen-bearing free radicals can—and do— tear apart the molecules within our cells.

Cancer, arthritis, and cataracts are among the degenerative ailments caused (at least in part) by free radicals. Those systems in your body that have the heaviest demand for oxygen are also the most susceptible to damage from free radicals: the brain, the heart, the skeletal muscles that move your limbs.

While the heart and skeletal muscles obviously work hard and demand oxygen, it may come as a surprise that the brain—which comprises only 2 percent of the body's weight—uses some 20 percent of the oxygen the body breathes in. The brain works hard, and it is *always* at work. This is why we mop our foreheads even when the rest of our body is not noticeably sweating. The brain's constant burning of oxygen produces considerable heat.

Oxygen radicals are particularly dangerous to the *mitochondria*, which are the tiny power plants within each

23

FREE RADICALS

Political radicals can be dangerous; molecular radicals can be killers.

In the microminiature world inside living cells, a free radical is a molecule in which one atom has gained an extra electron. This makes the molecule highly unstable and chemically very reactive. It wants to combine with another atom or molecule and get rid of that extra electron at all costs. It's an accident waiting to happen on the molecular scale.

The force that binds atoms together to form molecules is electrical. It is also the force that tears molecules apart. It is this electromagnetic force that makes chemistry work. Living cells are chemical factories.

Normally, atoms are electrically neutral. The positive electrical charges of their nuclei are balanced exactly by the negative electrical charges of the electrons orbiting around the nuclei. When an atom is stripped of one or more of its orbital electrons, it is called an *ion*, and it has a net positive electrical charge. Alternatively, an atom may gain an extra electron and thereby have a net negative electrical charge.

When a molecule contains one or more ions, it is a radical. This electrical imbalance makes radicals extremely reactive, eager to combine with another molecule. Biologists use the term *free radical* to denote electrically unbalanced molecules that attack the molecules in living cells.

cell that "burn" food and oxygen to generate energy. Damage to the mitochondria, of course, leads to lower energy production in the cell. As energy production declines, the cell begins to behave erratically, and this can lead to the production of still more oxygen-bearing free radicals. A vicious circle indeed.

Damage to the mitochondria accumulates throughout life, and this is now suspected as one of the leading causes of the effects we attribute to aging. We grow more feeble as our mitochondria crumble under the constant attack of oxygen-rich free radicals.

Moreover, oxidants such as hydrogen peroxide (generated naturally inside the body, not from hair bleach) can get into the cell's nucleus and damage the genes in the nuclear DNA. This may be a primal cause of cancer. Biologists have calculated that oxygen radicals damage the DNA in our cells some ten thousand times each day. Most of this damage is repaired by the cells' defense mechanisms, but the small amount that is not repaired accumulates in the cell day by day, year by year.

The body has natural defenses against free radicals: proteins produced by the cells that attack and destroy oxidant molecules. There is a bewildering variety of natural antioxidants, including uric acid, vitamin E, catalase, superoxide dismutase, glutathione, and glutathione peroxidase.

We can also get antioxidants from the foods we eat or from vitamin supplements. Vitamins A, C, and E can be especially helpful in tempering the molecular ferocity of oxidizing free radicals. But there is no reliable evidence that antioxidant vitamins, even in megadoses, do anything to retard the effects of aging.

Another source of oxidants comes from *macrophages*, the form of white blood cell that kills invading bacteria, parasites, or cells infected by viruses. A macrophage can digest an alien bacterium in less than a hundredth of a second! They engulf their prey and destroy it with powerful chemicals called *enzymes*. Unfortunately, oxidizing free radicals are produced as a by-product of the macrophage's work. But the story is not all bad. Most of the

25

free radicals are quickly destroyed in the blood by enzymes such as *catalase*.

Most but not all.

Nothing in the human body is simple. The paradox remains: Oxygen, essential for life, is one of the major causes of damage to our cells. Oxidation by free radicals is one of the ongoing processes of aging.

Sugar, Browning, and AGE

The other aging process stems from sugar.

Glucose is a simple sugar that is fundamental to the energy processes in our bodies. In our cells' energy-producing technique of burning food with oxygen, glucose is the food.

Glucose is produced by plants through *photosynthesis*. Green plants transform carbon dioxide from the air, plus water and sunlight, into glucose and oxygen. The oxygen is released into the air. For plants, it is a waste product; animals breathe it.

Animals also eat plants and convert their simple glucose into more complex molecules. In humans, the complex carbohydrates that we eat are broken down by our digestive systems into (you guessed it) glucose molecules.

Animals such as ourselves essentially burn glucose and oxygen to produce their energy of life. The process of *glycosylation* is an important and necessary part of cellular chemistry; it helps to convert proteins newly produced in the cell into functional proteins, able to take part in the cell's ongoing processes.

But just as the oxygen we need for life causes cellular damage, unfortunately, so does the glucose.

In 1912 the French chemist Louis Maillard discovered that when glucose is mixed with protein and heated, the liquified mixture changes color from clear to yellow to brown. You can see the same change when you cook meat: It turns brown.

So do our cells.

Patients with uncontrolled diabetes seem to display the effects of accelerated aging. Cataracts, atherosclerosis, heart attacks, strokes, stiffening of the joints and lungs all appear earlier in diabetics than in nondiabetics. By the 1970s, medical researchers found that uncontrolled diabetics have high levels of glucose in their blood and also have glucose attached to their *hemoglobin*, the blood protein that carries oxygen to the cells.

In the late 1980s, Anthony Cerami, a biochemist at Rockefeller University, suggested that the so-called "Maillard effect" (or browning) slowly accumulates in the body's cells, and this is one of the causes of aging. Browning attaches glucose molecules to certain proteins in our tissues. Cerami astutely dubbed these altered proteins *AGE*, an acronym for *A*dvanced *G*lycosylation *E*ndproducts.

One of the proteins that glucose turns into AGE is *collagen*. In youth, collagen is a springy, flexible strand, like a rope or a cable, that forms the structural foundation for the skin, lungs, and blood vessels. Collagen makes tendons and ligaments that can bend and twist without breaking. It forms the resilient cartilage that pads and protects joints such as the knees.

When collagen starts to brown, it stiffens. As glucose molecules attach themselves to the collagen, they form cross-links between the collagen molecules. The collagen strands become stiffer, harder to bend. We feel the results as, over the years, it becomes more and more difficult to bend, to run, or even to stand up straight without a grunting effort.

Collagen is also deposited along the interior walls of arteries, along with calcium, as we age. This stiffens the arteries, narrows their inner channels, and plays a role in the formation of cholesterol plaques that narrow the arterial channels even more.

Worse, AGE acts as traps for other molecules, including cholesterol-rich low-density lipids (LDLs) that clog arteries and lead to atherosclerosis.

Protein enzymes responsible for regulating the cells' activity, for activating some genes and suppressing others, are also affected by *glucose browning*. Such AGEd enzymes fail in these functions, leading to further damage to the cells. Moreover, glucose can bind directly to DNA; over time, the accumulating AGE can disrupt the DNA's proper functioning, upset the production of new cellular proteins, and lead to mutations.

And still worse, when such sugars attach to protein molecules, it makes the resulting molecule less soluble, more likely to solidify and become an inert solid chunk. Such solidified masses have been found in the brain lesions of Alzheimer's victims, where they are called *tangles* and *plaques*.

To date, the rate of browning—that is, the rate at which glucose forms AGEs in the body—has not been shown to correlate directly with the rate of aging. The link between the two is subtler than a straight one-to-one relationship. This probably means that the body has natural defenses against browning, anti-AGEing defenses. What they might be is currently unknown.

Aging and Reproduction

As we have seen, since long before writing was invented, much of the interest in aging came from men who dreaded the loss of their sexual prowess with advancing years. It is very common for a man to equate his social standing and personal power with his sexual capabilities.

Hence the desperate search for the secret of eternal youth, from the time of Gilgamesh and even before.

One route to an extended life span, paradoxically, is castration. *Testosterone*, the male sex hormone, suppresses the immune system. When the immune system is prevented from working properly, we become more susceptible to infectious diseases, from the common cold to AIDS and everything in between. Castration removes the *testes*, which are the glands that produce testosterone, and thus removes testosterone's suppression of the immune system.[3]

29

Castrated guinea pigs resist infections better than their testosterone-producing brethren. Castration also removes the risk of prostate cancer from adolescent dogs.

There is evidence that castration extends human life spans as well. Before the 1960s, it was not uncommon to castrate mentally retarded men who were institutionalized. A study conducted among the inmates in a mental institution in Kansas showed that castrated men lived an average of fourteen years longer than noncastrated. The younger the inmate was when castrated, the longer he lived. Few men, however, will volunteer for such a procedure in order to extend their lives a dozen years or so.

The male decline in sexual prowess is quite gradual. Men continue to manufacture sperm through most of their lives. Production of testosterone, the major male sex hormone, decreases about 1 percent per year, on average, beginning usually in the late teens. It is not unusual for men in their sixties to father children. And it is not unheard of for a man in his seventies to become a father.

Women face a much sharper reproductive limit. While men's sexual powers gradually ebb over the years, women

---

[3]Incidentally, stress also suppresses the immune system. Persons experiencing high levels of stress are more susceptible to infectious diseases than those whose stress levels are lower.

experience menopause when they have reached the end of their reproductive lives. In this sense, menopause is the clearest and sharpest example of aging that we know.

What can we learn from it?

To begin with, it is known that while high doses of *estrogen* (the major female sex hormone) suppress the immune system just as testosterone does, low estrogen levels can stimulate the immune system. Women are subject to high estrogen levels during pregnancy.

Female dogs are stricken with breast cancer at more than twice the rate of humans. If a bitch is spayed before the onset of puberty, however, her chances of contracting breast cancer go down to less than 1 percent of an unspayed female, because her main source of estrogen has been removed. If a female dog is spayed after having borne two litters of pups, her risk of breast cancer goes down 26 percent.

Breast cancer now strikes roughly 10 percent of American women. Before menopause, it seems clear that high levels of estrogen are one of the major causes of breast cancer. After menopause, when estrogen levels are greatly reduced, the risk of breast cancer is linked to obesity, because fat cells can convert other hormones and steroids into estrogen. Thus overweight women are more at risk for breast cancer after menopause than thinner women.

However, menopause generally slows the risk of breast, uterine, and ovarian cancers. Whereas before menopause the risk of contracting such gynecological cancers doubles each year for women in the United States, after menopause it takes thirteen years for the risk to double. Of course, there are many other factors involved, including environmental influences such as smoking; the given risk factors are averages for the population as a whole.

Breast cancer rates have been rising in the United States at a rate of about 1 percent per year. This can be correlated with the lowering of the average age of puberty and the increase of women's average age at menopause. Thanks to good nutrition and better environmental and social conditions in general, American women undergo puberty younger than earlier generations and remain fertile longer. Thus they are subject to high estrogen levels longer. One undesirable side effect of this is the increasing rate of breast cancer.

Menopause also has an effect on a woman's risk of heart disease. During their fertile years, women tend to have lower cholesterol levels than men, and the cholesterol they do have has higher levels of HDL (high-density lipoproteins, the "good" cholesterol) and lower levels of LDL (the "bad" cholesterol). During and after menopause, women's cholesterol profiles begin to look more like men's. Both total cholesterol and LDL levels increase—and so does their risk of atherosclerosis and heart attack and stroke.

Perhaps the most easily discernable effect of the loss of estrogen at menopause is *osteoporosis*, the thinning of the bones that leads to the stooped "widow's hump" posture. Bone tissue is just as dynamic as the rest of the body. It grows, reacts to stress, rebuilds lost or damaged cells, and heals when fractured. There is a constant give-and-take of calcium in the bone cells; calcium is lost through some biological processes and replaced through others.

One of the problems of long-duration space flight is that the bones slow their uptake of strengthening calcium under zero-gravity conditions, an apparently natural reaction to the lack of physical stress when the body experiences weightlessness.

Bones reach their maximum strength in the early adult years. They begin to slowly weaken in middle age, less so for men than for women, possibly because pregnancies take a good deal of calcium from the bones and teeth. Of course, calcium loss and bone weakening varies from one individual to another as a result of the amount of exercise they do, the nutritional levels of their diets, and so on.

During her menopausal years, a woman may lose 2 to 3 percent of the mineral mass of her bones every year. Once menopause is complete, the mineral mass loss settles down to about 1 percent per year. Osteoporosis begins when the bones start to lose more calcium than they can replace. The calcium is excreted through the urine. The bones become thinner and more fragile. Brittle bones are more easily broken, and fractured wrists and hips become common.

The "widow's hump" stems from the weakening of the spinal vertebrae, which no longer have the structural strength to hold the neck and back erect.

Hormone Replacement Therapy

These effects of menopause have become a problem only during the past couple of centuries. In earlier generations, very few women lived past their menopausal years.

If these effects are caused by the loss of hormones such as estrogen and progesterone, can life be extended—or at least made healthier and more tolerable—by taking doses of hormones to replace what has been lost through menopause?

Hormone replacement therapy (HRT) is relatively new, and its results are not yet very clear. While bioscientists can experiment on animals, they cannot do so on humans. Moreover, humans live seven or eight decades or even more. Lab mice can produce several generations in a few years; fruit flies in a few months. It takes a *long* time for studies of human biology to produce results.

Harvard University's ongoing Nurses' Health Study has since 1976 been tracking the health of fifty thousand nurses who started using estrogen replacement soon after menopause. During the first ten years of this HRT, their annual death rate was 40 to 45 percent lower than the death rate of women who did not use estrogen replacement; their risk of dying from heart attack was 53 percent lower. After ten years, the death rate among the HRT recipients dropped to "only" 20 percent lower than women who did not use hormone replacement therapy, primarily because of a 43 percent increase of death from breast cancer among the HRT users.

33

This is no Fountain of Youth, but it does extend the life span, mainly by reducing blood cholesterol levels to the point where they were before menopause and thereby reducing the chances of heart attack. There are other factors at work as well, including HRT's effect on body fat and blood pressure.

HRT recipients also showed far fewer bone fractures than nonusers, evidence that hormone replacement helps to slow bone mineral loss. Again, such results came from women who began HRT within a few years of the end of menopause and continued the therapy.

HRT apparently increases the risk of breast cancer, although some researchers maintain that by regulating dosages carefully, the risks of breast, uterine, or ovarian cancer could be brought down to much lower levels.

Hormone replacement therapy is so new, especially in comparison to the human life span, that the unknowns are many and deep.

However, as bioscientists learn more, HRT may become a routine part of their tool kit for extending women's life spans far beyond today's marks. Recognize, though, that if HRT turns out to be useful in extending women's life spans, it will be because the therapy reduces women's risks of heart attack and osteoporosis-related injuries. Hormone replacement will probably have no direct effect on aging itself.

Ovarian Transplants

While the search for the Fountain of Youth has been historically a male pursuit, modern scientific research is allowing women to join the quest as well.

The female reproductive system ages faster than any other part of the human body. Although a human female fetus starts with some 7 million eggs, all but 1 or 2 million of them die even before birth. Of these, only about 250,000 remain by the time of puberty. During each menstrual cycle, one egg will fully develop while hundreds of others quietly die. By the age of 35, this loss rate doubles, until by age 50, typically, a woman has depleted the store of eggs she was born with and undergoes menopause.

Yet in 1997 Arceli Keh of California gave birth to a healthy baby at the age of 63, thanks to artificial insemination with her 60-year-old husband's sperm and an egg donated by a younger woman and transplanted into her uterus.

Roger Gosden, professor of reproductive biology at the University of Leeds, has gained some notoriety in the news media by his efforts to reverse the effects of menopause through transplanting youthful eggs from a fetus or a young cadaver into the ovaries of older women.

"I never imagined myself being cast in the role of a dangerous radical," Gosden writes. "Lurid headlines about my attempts at Edinburgh University to reset the biological clock in women ushered in the 1994 New Year. . . . Dr. Frankenstein, it was said, was alive and working in Edinburgh!"

Gosden believes that it might be possible to rejuvenate the reproductive system of older women by transplanting youthful eggs into a woman's ovaries, thereby "putting menopause into reverse."

Although it is still too early to assess the results of Gosden's pioneering work, his experiments show that women are also interested in reversing the effects of aging and having babies well after the age when menopause normally shuts down the female reproductive system.

Down Among the Cells

It is clear that the effects of aging happen at the cellular level in our bodies. Something like the old aphorism "You are what you eat" is at work here: We are what our cells are. The 100 trillion or so cells that make up our bodies are constantly building, tearing down, producing proteins and enzymes, resisting infections—and dying.

To understand how our bodies work and how death may be averted, we must understand how cells work.

3

Cellular Life

Each of us is an ecosystem of unseen organisms.
— J. John Sepkoski, Jr.

∞ AGING TAKES PLACE WITHIN THE CELLS OF YOUR BODY. To understand aging and death and counteract them, we must understand how the cells work.

The human body is composed of some 100 trillion individual cells: nerve cells, muscle cells, skin cells, heart and liver and stomach cells, bone cells, hair cells—100 trillion individual units which, together, comprise *you*[4].

One hundred trillion cells. There are more cells in your body than there are stars in the entire Milky Way galaxy, by a factor of a thousand.

Cells are the smallest things that can be said to be alive. They have been around for a long time, almost since the moment our world first formed a solid crust.

The Earth was born about 4.5 billion years ago, coalescing out of the same cloud of gas and cosmic dust that

---

[4] A trillion is a million times a million. In scientific notation, 100 is 10^2; 1,000 is 10^3; 1 million is 10^6; 1 billion is 10^9; 1 trillion is 10^{12}; and 100 trillion is 10^{14}.

formed the Sun, planets, and smaller bodies of the solar system. At first the Earth was molten hot, incandescent from the constant pounding of infalling meteoroids and comets. We can see the results of that cosmic bombardment on the battered face of the Moon.

Our planet's winds and rains erased most of those primeval scars, but occasional meteoroids still streak down on us. A six-mile-wide rock slammed into the area that is now Yucatán about 65 million years ago. The fall-out from its impact wiped out three-fourths of all the life-forms on Earth, including the dinosaurs. About fifty thousand years ago, a much smaller stone, less than two hundred feet wide, dug out the mile-wide Meteor Crater in Winslow, Arizona.

Bacteria: The "Firstest and the Mostest"

Between 3 and 4 billion years ago, the surface of the newly born Earth cooled enough from its molten state to form a solid crust. Living cells arose in an eyeblink of time, geologically speaking. They became the first primitive bacteria.

For most of the Earth's history, life remained single-celled. There were bacteria and, eons later, other single-celled organisms. Nothing else appeared for at least 2 billion years. It was not until less than a billion years ago that the first multicellular organisms arose. The earliest fossils that are clearly multicellular date from about 750 million years ago.

Today the bacteria are still very much with us, the oldest and most abundant form of life on the planet. Paleoarchaeologist Stephen Jay Gould and other biologists estimate that the mass of bacteria living today is equal to the mass of all the other forms of life on Earth—from amebas to humans—combined.

Microorganisms and cells in general were unknown to humans until the mid-1600s, when Antonie van Leeuwenhoek of Holland and others began to use the newly developed microscope to study living matter. When they examined samples of ditch water, plant fluids, human saliva, and other liquids, their microscopes revealed swarming, teeming populations of what Leeuwenhoek called *animalcules:* little living things.

English scientist Robert Hooke popularized the term *cells* in his book *Micrographia*, published in 1665, comparing the microscopic units comprising the structure of cork to the individual little monks' cells in a monastery.

Immortal Bacteria

Bacterial cells are essentially immortal. They do not die of old age. They can be destroyed by ionizing radiation or extremes of heat or cold, or by being gobbled up by another creature, or by starvation or lack of water or other accidents or catastrophes from outside their own cell walls. But they do not age. If they can avoid the myriad dangers of the environment in which they live, they go right on living (see Radiation, page 39).

Bacteria can be killed, but they do not die at the end of a certain life span. They do not grow old and wither. Their life span, in effect, is infinite—although their life expectancy is certainly not.

It is possible (though extremely unlikely) that there are bacteria alive today that are millions, even billions of years old.

A bacterium reproduces by fissioning. It splits into two bacteria. Mother and daughter cell go their separate ways, fissioning again and again. They will not stop until they are killed by outside forces. A typical bacterium of

RADIATION

The term *radiation* is used in two ways, and it is important to understand the difference between them.

Light, in its many forms, is *electromagnetic radiation*. Visible light is only one small slice of the broad electromagnetic spectrum, the part that our eyes have adapted to sense. Light from the Sun is essential for virtually all forms of life on Earth, especially for chlorophyllic green plants and the creatures who live by eating them—including us.

Invisible to us are other forms of electromagnetic radiation: radio waves, microwaves, infrared light, ultraviolet light, X rays, and gamma rays.

The kind of radiation that can cause damage to living tissue is called *ionizing radiation,* because such radiation can ionize atoms within cells, strip electrons from those atoms, and thereby cause harmful physiological effects. Ultraviolet light, X rays, and gamma rays can harm cells in that way. These high-energy forms of electromagnetic radiation are also dangerous forms of ionizing radiation.

Other forms of ionizing radiation come from energetic subatomic particles, such as those given off by radioactive elements.

about one cubic micrometer[5] in volume—such as the rod-shaped bacillus that causes bubonic plague—can divide in as little as twenty minutes. The *Escherichia coli* that dwell in your digestive tract double their numbers every fifteen minutes in laboratory cultures.

---

[5]One micrometer is one millionth of a meter or slightly less than four ten-millionths of an inch. Two hundred thousand bacteria could be sitting on the dot at the end of this sentence.

Still, this means that in roughly two days the daughter cells of just one little *E. coli* would weigh as much as the entire human race. A couple of days later the descendants of that one bacterium would outweigh every other living thing on Earth combined—if each cell survived.

If those outside forces of heat, cold, radiation, starvation, and hungry predators (including other bacteria) did not constantly cut down the bacterial population, the Earth would have been buried by bacteria long ago, because, left on their own with plenty of food and no outside threats, they would continue to thrive and reproduce forever. As we will soon see, human cells are not so prolific—except for tumors.

For untold eons, bacteria were the only form of life on Earth. Today we tend to regard them as primitive—or even worse, as dangerous "germs." They are far from primitive, as Appendix A: Bacteria shows. But some of them, such as the plague bacillus or the pneumococcus that causes pneumonia, can be quite dangerous indeed.

Prokaryote vs. Eukaryote

Life evolved into larger single-celled creatures: amebas, diatoms, paramecia, algae. A typical paramecium, for example, is thousands of times bigger than a typical bacterium. Even so, it is still microscopically small.

One of the major differences between these single-celled *protists* (as scientists call them) and the earlier bacteria is that the protists have a nucleus in their cells, while the bacteria do not.

Bacterial cells are called *prokaryotes* (from Greek, meaning "before the seed" or "before the nucleus"). Their genetic material, the DNA that carries the blueprint of the cell's structure, is strung along a single, threadlike *chromosome* that is curled into a ragged loop that floats in the cell's *cytoplasm*, the protein-rich liquid that fills the cell.

All other organisms—including humans—are *eukaryotes* ("good seed"); each of their cells has a nucleus in which the DNA is arranged in many long X-shaped chromosomes. The nucleus is walled off from the rest of the cell, but the wall membrane is studded with pores that allow selected molecules in and out.

Eukaryotic cells are filled with specialized units for producing energy, manufacturing new proteins, and getting rid of wastes. While each type of specialized unit has a specific name of its own, biologists have a catch-all name for them; they are called *organelles*.

Eukaryotic cells can also die from aging, unlike the immortal prokaryotes. The cells of our bodies are subject to senescence—aging. They live and reproduce for a certain span, then they wither and die.

At least, most of them do. Tumor cells do not age; they are as immortal as the bacteria.

Haploid vs. Diploid

Eukaryotes have another important difference from prokaryotes; they carry two copies of each chromosome inside their nuclei, one set from each parent. Prokaryotes have only one set of genes; they are called *haploids*, from a Greek root meaning "single." Eukaryotes, including us, are *diploids* and have two sets of gene-bearing chromosomes.

Every cell in the human body is diploidal, except for the male sperm cells and the female egg cells, which are haploidal. They unite their chromosomes when the ovum is fertilized by a sperm and become diploidal.

Being diploid offers a tremendous advantage. Since the chromosomes carry the cell's genes, its genetic blueprint, it is enormously helpful to have a backup copy ready to use if the original is lost or damaged.

In a diploidal cell, if one copy of a gene on a certain chromosome is defective, the other copy on the matching chromosome may be able to do the required job, and the cell can work as if nothing is wrong. The gene that is actually working, or *activated*, is called *dominant*; its silent, unactivated partner is a *recessive* gene. The Austrian monk Gregor Mendel discovered in the 1850s how the diploidal genes of pea plants hand down their dominant and recessive traits from one generation to the next, even though he had no idea of what genes were. His work, unnoticed for almost fifty years, laid the foundation for the modern study of genetics.

Sex and the Diploidal Cell

Being diploidal also conveys an even more profound benefit. Diploidal organisms can shuffle genes, exchange chromosomes, and thereby change their genetic blueprint. This is what we know as sexual reproduction. And, very much as the Biblical tale of Genesis tells it, with sex came death. Fissioning bacteria may be immortal; creatures that have sex are not.

Yet sex conveyed such enormous benefits to the protists that it became the standard method of reproduction for almost all the multicellular creatures that followed them. That ability to shuffle genes around was worth dying for, as far as nature was concerned.

When such an exchange of genes results in an improved organism (for example, more likely to evade predators, more resistant to disease, or more efficient in metabolizing food), that organism prospers. When the change is deleterious, the corresponding set of genes on the matching chromosome may be able to take over and save the creature from extinction.[6]

At the tip of each spindle-shaped chromosome is a sort of cap, called a *telomere*. Telomeres somewhat resemble the aglets on the ends of shoelaces. The telomeres keep the ends of the chromosomes from sticking together and from sticking onto other chromosomes. Bacterial DNA does not have telomere caps and tends to loop itself into a ragged circle, like a snake swallowing its tail.

43

Telomeres keep the individual strands of DNA in eukaryotic cells from looping or connecting to one another. They also play an intriguing role in cellular aging. Some researchers believe that telomeres are a sort of cellular clock that sets the rate at which the cells age and eventually die.

We will see much more of telomeres in later chapters.

Sex and Death

The first eukaryotes were single-celled creatures, such as the paramecium. They invented sex.[7] They also invented aging—and death from old age.

Biologists have found that if they put a single paramecium in a laboratory culture dish with plenty of food and room, it starts merrily reproducing by fission, just as bacteria

---

[6]I use the term *matching chromosome* rather glibly here. In nature, the pairs of chromosomes are seldom perfect mirror images of one another. Still, diploidy is so advantageous that once it was developed, all succeeding organisms used it.

[7]That may be an overstatement. Increasing evidence shows that some bacteria exchange genetic material through *transduction*, where viruses transfer genes from one bacterium to another. Other bacteria transfer plasmid circles of DNA through conjugation.

do. But after a while, the rate of division begins to slow down. After about two hundred cell divisions, the paramecia will stop dividing and will die. Unless they have sex.

Sex alters the picture.

Paramecia *conjugate*. That is, two paramecia will meet, stick together briefly, and exchange parts of their chromosomes. I still remember the shock I felt when, on our first day of using microscopes in high school biology lab, one of my buddies cried out, "Hey! I got two parameciums screwing!"

When a paramecium conjugates, it renews the creature's vigor. After conjugation, it begins to reproduce by fission once again just as rapidly as it did when it was "young." Somehow conjugation—the earliest form of sex—reverses the effects of aging. While conjugation does not directly cause reproduction, it allows the paramecium to begin fissioning again, just as it did in the beginning. For paramecia and many other protists, sex and reproduction are separate (though related) functions.

Biologists use the term *senescence*, which is defined as the progressive deterioration of body functions over time: the process of aging. For human beings, aging begins around 10 or 11 years of age, slightly before the onset of puberty.

Paramecia that do not conjugate become senescent and eventually die. It is as if there is a clock ticking away somewhere inside that single cell, and when it counts down to zero, the creature dies. But if the paramecium can find another paramecium to conjugate with, the senescence clock is reset. It becomes young and vigorous again.

Sex vs. Cloning

Sex was a profound invention, genetically speaking. When bacteria reproduce asexually, through fissioning, they are making clone copies of themselves. The daughter cell has exactly the same genetic makeup as the mother cell. If, during the bacterium's lifetime, the DNA of its genes is somehow damaged—by ionizing radiation, for example—the bacterium passes on an exact copy of the altered DNA to its offspring. Such *mutations* (alterations of the genes) can be harmful and result in a creature that cannot survive as well as its original DNA would allow. Thus asexual reproduction—fissioning—reproduces "bad" DNA, just as exactly as it reproduces "good" DNA.

45

When paramecia conjugate, however, they exchange parts of their chromosomes, which are the stringlike structures that contain their DNA. Neither paramecium is the same as it was before conjugation, genetically speaking. Its genes, which reside on the chromosomes, have been changed. Think of it as swapping clothes with a friend. You meet and exchange, let's say, the sweaters you are wearing. And your shoes. Then you go your separate ways, each of you dressed a bit differently than you were before you traded clothing.

That is what happens to a paramecium's genes when it conjugates. It has exchanged some of its genetic material for some of the genetic material of its conjugation partner. When the paramecium reproduces again, by fissioning (the only way paramecia reproduce), its daughter is somewhat different than she would have been if mother had not conjugated first.

Sex allows creatures to alter their genetic makeup. Instead of identical clone copies, the offspring are a bit different from the parent. This is an incredibly important advantage in a world where the environment is constantly changing. If every member of the family is exactly the same, genetically, then an environmental shift that can kill one member of the family can wipe out the entire lineage. Mixing up those genes through sex gives a species some hope of surviving environmental shifts that would push some members of the clan over the brink of extinction.

Sex is such a useful trick, genetically speaking, that every species of creature on Earth—other than the bacteria—uses it.

But if sex is so great, why do we die? How did it happen that sex, which rejuvenates protists, does not rejuvenate us? Why does senescence overtake us after our "threescore years and ten . . . [or] forescore" years?

Genes Beget Genes

Which came first, the chicken or the egg? To geneticists, it was definitely the egg. It is an old saw among genetic researchers that the human body is merely a gene's way of making another gene.

Looked at dispassionately, the "purpose" of any organism is to reproduce itself. Bacteria do it quite simply: One bacterium splits into two, each cell with identical genes. Paramecia and other protists fission as well, although paramecia become senescent and die unless they can conjugate.

Multicellular animals reproduce sexually. They mate. And die. Once the organism—a fruit fly, a tropical orchid, a frog, an azalea, a human being—has lived long enough to reproduce, it begins to senesce and eventually it dies. It makes no difference if the organism has actually been successful at mating and reproducing. Its in-built biological clock begins ticking down toward death just as the organism reaches sexual maturity.

"Be fruitful and multiply" is nature's command. And once you have had the chance to obey that command, you are on the road that leads to death. It is as if the genes' only interest is in seeing themselves reproduced. Once sexual maturity is reached, once the organism has achieved the capability of reproducing its genes, the rest of the body is expendable. It can be thrown away, because the genes have (in most cases) been transmitted to a new generation.

In this view, everything about us—except our egg and sperm cells—is expendable, excess baggage that is needed only until we have reached the age where we can reproduce.

47

The Hayflick Limit

The American cell biologist Leonard Hayflick found experimental evidence that powerfully reinforces this concept. In the 1970s he began to study the reproduction of human *fibroblast cells*, the kind of cells that make connective tissue throughout the body and form scar tissue around a wound. Hayflick incubated fibroblast cells in a culture dish with plenty of nutrients and the same temperature, humidity, and mixture of gases, such as oxygen, that are found in the human body.

Most human cells are specialized and reproduce rarely or not at all, but fibroblast cells have the ability to fission again and again, very much like bacteria or protists, such as the paramecium.

Although the fibroblasts reproduced more slowly than paramecia, they quickly began dividing. But soon their rate of division slowed down. After a few weeks, they stopped altogether and eventually died.

Hayflick found that fibroblast cells from a human fetus would divide about fifty times, on average, before stopping. Fibroblast cells from a middle-aged person might go through twenty or thirty divisions; those from an elderly person would reproduce only a dozen or so times.

The most startling result of Hayflick's work was that when fetal fibroblast cells were stopped from reproducing after twenty doublings and frozen in liquid nitrogen—in some cases, for thirty years—after they were thawed and put back in a culture dish, they reproduced another thirty times before dying.

The *Hayflick limit*, as it has become known, is the number of times a cell will reproduce. Each type of cell has its own limit. For human fibroblast cells, the limit is about fifty cell divisions. Does the accumulated damage from free-radical oxidation and glucose browning bring the cells' ability to reproduce to an end? Or is there is a program written into the cells' genes that allows a fibroblast cell to reproduce a total of some fifty times—and that's all? Then the cells die.

48

The Immortal HeLa Cells

This is not so with tumor cells. In 1951 a 30-year-old African American woman, Henrietta Lacks, was admitted to Johns Hopkins Hospital in Baltimore, suffering from cervical cancer. A small slice of her tumor was removed for analysis. She was given radiation therapy, but she died eight months later.

Her tumor cells live on, however. Johns Hopkins research scientist George Gey acquired a sample of Ms. Lacks's tumor cells and put it into a culture dish, much as Hayflick would later do with fibroblast cells. Gey needed human cells that could be grown in the laboratory as part of his research on polio.

There was no Hayflick limit for the tumor cells. They multiplied and continued to multiply. No matter how many times they were thinned out, no matter how many samples of the *HeLa cells*[8] were sent to other researchers,

---

[8]Originally so named to protect Henrietta Lacks's privacy, even after death. The cells have become so useful and important in medical research, however, that she is now quite famous.

the cells kept on going and going, just like the pink bunny on the television battery commercials. Viruses grew in them very well, which delighted Gey and his colleagues, because they wanted to study how the polio virus attacks human cells.

Soon HeLa cells were being shipped to researchers all around the world. Some were even sent into space aboard the *Discoverer 17* satellite.

HeLa cells in culture double about once a day. They have been doing so for more than seventeen thousand days. They have escaped whatever internal clock there is in normal cells that sets the Hayflick limit.

Like bacteria, tumor cells are effectively immortal. They do not senesce and die of internal causes, although they can certainly be killed. Unless they are stopped by outside agents, tumor cells continue growing, which is what makes them deadly when they grow inside a human body.

If this type of cell can be immortal, what happens to normal human cells that results in death?

4

Cellular Death

We die because our cells die.

—WILLIAM R. CLARK

∞ IT IS TIME FOR US TO LOOK DEATH IN THE FACE.

Let us examine a single cell in a human body, a cell that can age and die, one of the cells in the muscle tissue of the heart, which is called a *myocardial cell.*

A Specialized Cell

Unlike the bacteria and other single-celled creatures, the cells of multicellular organisms such as ourselves have specialized functions. A myocardial cell's job is to help the heart pump blood through the cardiovascular system to all the other 100 trillion cells of the human body.

In the 1960s I worked on one of the early research programs to develop an artificial heart at the Avco Everett Research Laboratory in Massachusetts. At the beginning of the program, the physicists and engineers involved looked on the heart as a fairly simple pump, which should be replaceable by a similar pump made of plastic. The scientists and engineers soon found, however, that while the pumping action of the heart may be simple, the ability to change the rate at which it pumps is not.

Myocardial cells beat in response to very subtle chemical and electrical messages from the nerves that control the heartbeat. These messages can change in seconds or less in response to physical exertion or emotional stress. While scientists and engineers can build mechanical pumps that equal the natural heart's pumping power, they have yet to design sensors equal to those that the natural heart responds to.[9]

The myocardial cell spends its lifetime contracting and then relaxing, contracting and relaxing, in rhythm with all the other myocardial cells, sixty or seventy times each minute (and sometimes more), in response to the signals it receives from the nervous system. It is like a galley slave, pulling on his oar in response to the tempo of the galley master's drumbeat.

Because it is a specialized cell, with only one (very important) job to do, the myocardial cell does not move about independently. It does not go hunting for food, and it cannot protect itself against predators. Like a galley slave chained to his oar, the myocardial cell depends on others to bring food and drink to it. Virtually all the cells of our bodies are specialized and fixed in place.

Blood and Lymph

It is the bloodstream that carries life-giving oxygen and nutrients to all the cells of the body and carries away the waste

---

[9]There is also the vital matter of designing the artificial heart so that blood does not clot inside it. Clotting is a major cause of failure in artificial heart experiments.

products—such as carbon dioxide—that are the results of the cell's metabolism. Starting from the heart itself, the arteries branch into smaller and smaller vessels, down to microscopic-sized capillaries, to reach every cell in the body.

Each cell is bathed in a fluid derived from the blood, called *lymph*. Similar in composition to blood plasma, lymph bears the type of white corpuscles called *lymphocytes*.[10] Lymph seeps through the tiniest capillaries into the minuscule spaces between the cells. All cells live in a liquid environment, be they bacterial or human. This is strong evidence that life began in the water of primordial seas or ponds.

The myocardial cell pumps day and night, year after year, never tiring, constantly rebuilding itself—as long as it receives the oxygen and nutrients it needs. Its internal workings are capable of repairing damage to the cell to a certain extent.

Inside the oblong-shaped cell are sheets of protein that are attached to each end of the cell. They can contract and then spring back to their original lengths, rather like bungee cords. When the nerves controlling the heartbeat send out their signal, these protein sheets contract. If the heart is working properly, all the myocardial cells contract in proper rhythm and the heart beats normally, forcing blood into the aorta and eventually to every cell in the body—including the myocardial cells themselves.

The Inner Workings

The inner workings of the cell is a complex, frantic world of molecules: long strings of atoms that are coiled into

---

[10]White blood cells are called *leukocytes*. They come in two major varieties: *macrophages* and *lymphocytes*. There are two types of lymphocytes, named after the tissues that produce them: *B cells* come from the bone marrow; *T cells* mature in the thymus.

compact shapes, rather like twisted and folded Slinky toys. These organic molecules are all based on chains of carbon atoms, although there are plenty of other kinds of atoms connected to them: nitrogen, hydrogen, potassium, and many others.

The cell is like a frenetically busy factory, with molecules zipping back and forth constantly. Things happen in seconds here, even fractions of a second. Nothing stands still; the cell is constantly building new molecules and tearing down old ones. It is a dynamic, living organism. Only when it dies does it become still and quiet.

The myocardial cell has a nucleus that contains its *chromosomes*, the long stringlike structures made up of the DNA that composes the genes. DNA is a huge molecule; it contains millions of atoms. As we will see in the next chapter, its physical form is a double helix, like a twining spiral staircase. Strung along the DNA's double spiral are subunits called *genes*. The genes hold the blueprint of the cell's structure.

Coiled up molecules of *messenger RNA* (mRNA) are constantly passing through pores in the outer envelope of the nucleus, like messengers rushing in and out of a factory director's office. Within the nucleus, these molecules are given instructions for producing new protein molecules. They leave the nucleus carrying the blueprint information for the proteins that the genes have impressed on them. Again, like a factory, the messengers (mRNA) are given instructions by the factory's director (the genes) as to what products need to be manufactured.

Organelles

Outside the nucleus, the cell contains many organelles, each with a specific job to do. The *mitochondria* are the cell's power plants (see Marriage of the Mitochondria),

53

the *ribosomes* are its assembly shops, and the *lysosomes* are the garbage removal units.

The ribosomes take the information carried by the mRNA molecules and use it to build the many different

MARRIAGE OF THE MITOCHONDRIA

Between 2 and 3 billion years ago, roughly, a microscopic marriage took place. You carry the results of that marriage in every cell of your body. In fact, every eukaryotic cell on Earth bears the fruit of that marriage.

The earliest bacterialike organisms got their energy from sulfur and other elements, not from oxygen. To them, oxygen was a deadly poisonous gas. Their descendants are still among us, living in swamps and hot geysers and deep underground; they are the *anaerobic bacteria*. Oxygen kills them.

Some of those early organisms, however, produced oxygen as a waste product. The waste gas began to accumulate in the Earth's atmosphere.

Oxygen is a very energetic element, and before long some bacterial organisms began to use oxygen as their energy source, combining it chemically with carbon and hydrogen to produce carbon dioxide and water, just as today's animals do.

Somewhere, somehow, a single-celled organism swallowed one of these oxygen-using bacteria. Instead of digesting it, though, the two organisms began to live together. The oxygen user not only absorbed the waste oxygen her host was producing but provided energy for the host cell as well. That marriage, or *symbiosis* (an interaction between two organisms that is beneficial to both of them), is still going strong.

The oxygen users are now called mitochondria, and they can be found in every eukaryotic cell, from yeast to human. The mitochondria are the cell's power generators, combining oxygen and glucose to produce *adenosine triphosphate* (ATP), the chemical that transfers energy from one part of the cell to another.

By utilizing mitochondria, eukaryotic cells are able to generate twenty times more energy than anaerobic cells can. Moreover, by soaking up oxygen inside the cell, the mitochondria reduce the damaging effects of oxidizing free radicals.

The mitochondria have their own DNA, separate from and much smaller than the DNA in the cell's nucleus. Like the DNA of bacteria, mitochondrial DNA is looped into a circle. Their genes carry the code for only thirteen proteins, whereas the chromosomal DNA in the cell's nucleus can produce some ten thousand proteins.

Mitochondria are inherited only from the mother's egg cells; sperm cells do not contribute to them. Therefore geneticists and anthropologists have been able to trace human ancestry back through time by noting differences in mitochondrial DNA. Such studies indicate that our species originated in Africa about 200,000 years ago.

proteins that make up the cell's structure. A typical human cell contains some ten thousand different proteins. The cell is constantly being rebuilt with new protein molecules replacing worn ones. The ribosomes are the cell's assembly shops, where the actual work of manufacturing the needed protein molecules is done.

While most of the proteins manufactured by the ribosomes are used to maintain the myocardial cell itself, some of the proteins thus manufactured are specialized molecules called *enzymes* that help the cell's various mechanisms to function properly. Remember, the myocardial cell's only purpose is to keep those muscle fibers contracting in rhythm to the heart's natural pacemaker signals. In essence, the cellular factory operates to keep the factory operating.

However, some cells in the human body produce proteins and steroids for export. For example, glands such as the thyroid and pituitary produce *hormones* that are pumped into the bloodstream. The pancreas makes insulin. The sex glands manufacture steroids. The ovaries produce estrogen; the testes produce testosterone.

Among the cell's organelles are garbage disposal units, called *lysosomes*. Waste products, molecules that cannot be used to generate energy or to manufacture new useful proteins, are dissolved into fragments of molecules by powerful enzymes and either carried outside the cell or recycled and combined into new, useful proteins.

The Cellular Membrane

The outer membrane of the cell is not the tough hide of a bacterium but a soft, spongy envelope composed mostly of fats and cholesterol. Like most of the cells in a multicellular organism, our myocardial cell does not need to face the harsh realities of the outside world; the realm inside the human body is maintained at a remarkably stable temperature and interior pressure. Most of the time.

The primary purpose of the cell's outer membrane is to keep the liquid environment inside the cell—its *cytoplasm*—separated from the rather different liquid environment on the outside. The cell's cytoplasm contains its own mixture of proteins and sodium, calcium and potassium ions at concentrations much higher than the lymph outside the cell; the lymph has a much higher concentration of water than the cell's cytoplasm.

The cell's outer membrane is not merely a passive bar-

56

rier, however. If it were, water from the lymph would tend to seep into the cell, causing it to swell and burst. So the membrane is lined with powerful molecular pumps. Some of them pump water out of the cell, back into the lymph, much as mechanical pumps in Holland are used to keep seawater from encroaching on land reclaimed from the ocean. Other pumps work away at maintaining the proper balance of ions inside the cell. If they stop, the cell quickly dies.

The myocardial cell is busy beating, every second of every minute of every hour of every year of your life, taking in nutrients which the mitochondria convert into energy, manufacturing new proteins in the ribosomes, while the lysosomes get rid of wastes.

The Way of Death

Now suppose the bloodstream that supplies the oxygen and nutrients to the cell is interrupted. The capillary that supplies the cell and its neighbors has become blocked by fatty deposits of plaque or a blood clot so that the normal flow of blood has become a tiny trickle. (Or if you are a detective story fan, imagine the capillary severed by a bullet.)

As the blood flow slows, the flow of lymph bathing the cell also dwindles. This means that the cell's vital supplies of oxygen and nutrients decrease to dangerously low levels. Inside the cell, the mitochondria— the energy generators—begin to shut down, just as an electricity-generating power plant will shut down when it runs out of fuel. The cell has backup emergency generators that burn starches and fats stored between the cells, but they cannot run for long. The cell will even start to consume some of its own proteins in a desperate effort to stay in business, but this too is only a stopgap.

Once the emergency energy generators run out of fuel, the inner workings of the cell begin to shut down. The contraction of the "bungee cords"—the real purpose of the cell's existence—ceases. The manufacture of new protein by the ribosomes stops. The DNA in the nucleus keeps functioning for a while, sending out messenger RNA molecules with instructions to manufacture new proteins, but the messages go unread because the ribosomes do not have the energy they need to build new proteins.

Meanwhile, the lysosomes are working overtime (as long as they have the energy to work at all), because the cell is filling up with unused and unusable molecular "garbage."

All through these desperate moments (we are talking about a few seconds), whatever little energy supplies that remain within the cell have been directed to the molecular pumps along the membrane wall. If they shut down, the cell will be destroyed. But eventually there is simply no energy supply left at all. The pumps quit at last. Calcium ions from the lymph slip through the membrane's pores and begin to destroy the now-silent mitochondria. And water rushes in, like the ocean pouring through the ripped hull of the *Titanic*.

The cell starts to swell, its membrane wall is stretched until it breaks, and the inner contents of the cell spill out into the lymph.

The cell has died.

If enough myocardial cells die like this, the heart they belong to is unable to pump blood adequately. The person to whom the heart belongs suffers a *myocardial infarction*: a heart attack. If the damage to the heart is bad enough, the person dies.

Necrotic Cell Death

The process described here is called *necrotic cell death,* or *necrosis.* It happens when the cells of the body are no longer able to function. Necrotic cell death comes through accident, or violence, or from toxins released inside the body by bacteria or other harmful microscopic organisms, such as the amebas that can cause fatal forms of dysentery. In necrotic cell death, the cell is usually torn apart, destroyed, by the flood of water pouring into it that rips open the cell's outer membrane.

When enough of our cells die, we die.

Yet there is another form of cell death that is much gentler and goes on inside our bodies all the time.

59

5

Programmed Cellular Death

The Leaves of Life keep falling one by one.
                              —EDWARD FITZGERALD,
                        The Rubáiyát of Omar Kayyám
                                          (Stanza 8)

∞ AT THIS VERY MOMENT, THOUSANDS OR EVEN MILLIONS
of cells in your body are committing suicide. It happens
all the time, even from before your birth, while you were
a fetus developing in your mother's womb.

This form of cell death is very different from necrosis.
Biologists have named it *apoptosis* (ap-oh-TOE-sis), from
the Greek word for the "falling of leaves from a tree" or
"the petals from a flower." As its name suggests,
apoptosis is a comparatively gentle form of death, quite
different from the violence of necrotic cell death.

Cellular suicide plays a role in many diseases. It may also
hold a key to extending human life span and immortality.

Programmed Cell Death

As the human fetus develops in its mother's womb, it
undergoes many transformations. At one point the fetus

has gill structures in its neck region; later on it shows a distinct tail. This led the German naturalist Ernst Haeckel (who, incidentally, coined the word *ecology*) to propose in the nineteenth century the concept of "Ontogeny recapitulates phylogeny." That is, the development of the fetus in the womb essentially replays all the biological forms of the fetus's evolutionary ancestors. While modern biologists believe Haeckel overstated the case, it is perfectly true that at various times during human gestation the fetus somewhat resembles an embryonic fish, tadpole, and so on.

61

The human fetus's limbs begin as small bumps at the end of the fourth week. Two weeks later the arms and legs are clearly discernable, although the hands look more like paddles than human hands. The feet, which develop a few days behind the hands, also have webbing between what will eventually be the toes.

The limbs of a developing fish keep this webbing and reinforce it to produce fins. Water birds such as ducks and pelicans keep the webbing between their toes, as do aquatic mammals such as otters.

Yet in the human fetus, between the forty-sixth and fifty-second day in the womb, the webbing between the fingers disappears, leaving a perfectly formed human hand. A few days later the toes also lose their webbing.

What happens to the cells that made up the webbing? Or the cells that composed the fetus's tail, earlier in its gestation?

They die.

They do not suffer necrotic cell death, however. They do not die because their blood supply has been cut off and they cannot get the nutrients they need to continue living. They do not die because some toxin from outside the cells kills them.

They commit suicide.

They die *apoptotically*, a form of cell death that is very

different from necrosis. They die because they are *programmed* to die. Something within those cells tells the cells to commit suicide, and the cells have no way to resist that command.

Imbedded somewhere in the cell's DNA is a genetic message that instructs the cell to die. This is called *programmed cell death* (PCD).

62

The Suicide Process

When a cell commits suicide, its DNA sends one final message to the organelles: the ribosomes, mitochondria, and lysosomes. Then the DNA starts to break up into tiny fragments, each too small to produce coherent instructions to the cell's machinery outside the nucleus. It is as if the cell's central office closes itself down and tears all the cellular blueprints to shreds.

Outside the nucleus, work goes on as normal—for a while. The cell's factory machinery chugs away, still faithfully fulfilling commands sent earlier by the DNA. Once the suicide message is read by the cell's machinery, however, the death process starts.

The cell detaches itself from its neighboring cells, disconnecting the contacts of its outer membrane from the membranes of the cells around it. Once the cell is alone, it begins to shudder, its outer membrane undulating like an untended sail flapping in the wind. The undulations become deeper, the membrane folding in on itself until small pieces of the cell—called *blebs*—break away and float off in the currents of the surrounding lymph.

These cell fragments are devoured by neighboring cells; in death, the cell provides nourishment for its neighbors. At last nothing is left of the cell. It has committed suicide and its mortal remains have been recycled to help other cells live.

Avoiding PCD

Programmed cell death is involved in many aspects of fetal development: for example, in the development of the human nervous system.

Nerve cells (called *neurons*) in the brain and spinal cord are connected to other cells in the body by long thin wirelike extensions called *dendrites* that carry electrical signals to the neuron from the other cell: a muscle cell, for example, or a sensory receptor, such as the cells in the skin that respond to touch. During gestation, the fetus's brain and spinal cord neurons send out tremendous numbers of these fibers, each of them seeking a connection with another cell. When such a fiber makes a connection—with a muscle cell, let's say—it becomes the communication link between the central nervous system and that particular muscle cell for life.

63

But if a blindly seeking neuronal fiber finds no other cell to connect to, it quietly commits suicide. It goes through programmed cell death.

Similarly, white blood cells (lymphocytes) constantly patrol the bloodstream and lymph seeking foreign invaders. When a lymphocyte finds a bacterium or a cell that has been invaded by a virus, it kills it. If it does not find an invader after a certain allotted time, it follows its internal instructions and commits suicide, while the body generates new lymphocytes to keep up the patrol.

But if a lymphocyte does encounter a foreign body, its suicide program is shunted aside. The lymphocyte can live for ten years or more, carrying with it the chemical memory of the invading microbe so that the immune system will recognize such an invader again. Eventually, though, if that particular pathogen does not show up in the body again, the lymphocyte will quietly kill itself by apoptosis.

Thus the body can acquire an immunity against certain pathogens but may lose that immunity over time. This is the basis for vaccinations and later booster shots. The original vaccination inserts a weakened form of a certain pathogen into the bloodstream, where lymphocytes find it and attack it. Then the lymphocytes carry the chemical "signature" of that pathogen for years, on guard against a fresh invasion. Over time, though, the lymphocytes will commit suicide unless more such invaders are found. Booster shots provide enough of the weakened pathogen to keep the lymphocytes alert and active.

The important point, however, is that there is some mechanism in the cells that can cancel the suicide program. Lymphocytes may be scheduled for death by apoptosis, but they can generate a reprieve from that death sentence—sometimes.

Skin and Eye Lenses

Apoptosis is going on in your body at this moment. Millions of your cells are committing suicide, sacrificing themselves for the greater good of your existence.

During a woman's menstrual cycle, the cells of the lining of the uterus slough off after they experience apoptotic demise. Every few days, the lining of your stomach sloughs off, too, to be replaced by fresh cells.

Skin cells called *keratinocytes* are generated in deep layers of the skin, then migrate toward the surface. Before they reach the surface of the skin, they commit suicide or, to use the technical term, they die apoptotically. The dead cells are not digested by their neighbors, however. Their inner contents—their cytoplasms—are replaced by the tough protein keratin and they acquire a water-repellant coating. These dead cells become the outermost protective layer of the skin, until they are eventually abraded away or sloughed off, to be replaced by new keratinocytes.

Within the paper-thin depth of the skin, a regular conveyor belt operation proceeds constantly, taking two to three weeks for a fresh cell to rise to the surface, commit suicide, and eventually be lost and replaced.

The lens of the eye is also made of apoptotic cells that have committed suicide and had their cytoplasms replaced by a protein called *crystallin*.

Apoptosis and the Hayflick Limit

If some of the cells of our bodies are programmed to commit suicide on command, do *all* our cells carry this "death program"? The answer appears to be: Yes.

Remember the Hayflick limit. Normal human fibroblast cells, kept in a culture with plenty of nutrients and everything they need to survive, will reproduce no more than about fifty times. Other types of cells also have their own Hayflick limits. Once they reach their limit, despite all the nutrients and tender loving care lavished on them, they curl up and die.

The question then becomes: Is the death of a human being caused by this death program in our cells? Is aging and eventual death caused by a genetic program that forces our cells to commit suicide? Remember, when enough of our cells die, we die. Is *my* death preprogrammed? Is there a command buried deep in the DNA of my cells that is eventually going to be triggered and result in my demise?

But remember also the humble paramecium. This single-celled organism grows senescent and dies—unless it can conjugate with another of its kind. For the paramecium, there is also an internal clock ticking away. Yet the act of exchanging DNA with another paramecium resets that clock and rejuvenates the organism.

Can we find ways to reset the internal clocks in our

cells (assuming they have such clocks) and prevent the death program from starting?

Finding the Trigger(s)

The search for the trigger (or triggers) that cause cells to commit suicide is going on in many laboratories around the world. But the search is not an easy one. In the words of immunologist Richard C. Duke of the University of Colorado, "Investigators have been especially stymied in finding the molecules that directly activate the process."

Nothing that happens within the human body is simple. For example, the genes direct the assembly of mRNA, the process called *transcription*. The mRNA carries this information to the ribosomes, where proteins are manufactured; this process is called *translation*. That much is straightforward enough.

Yet there are also *regulatory genes* that can either promote other genes to do their work or prohibit them from doing it, *repressor genes* that inhibit the translation of the genes they are associated with, and *activating genes* that expedite translation of their associated genes.

Writing about the way genes control the manufacture of proteins within the cell, neurobiologist Michael Fossel points out, "Like a biological Rube Goldberg contraption, gene A activates the translation of gene B, which in turn both activates translation of gene C and represses gene D translation. Gene C represses the expression of gene A (which started the whole intricate process), completing a feedback circuit and thereby shutting itself down again."

The Czech science fiction author Karel Čapek is famous for coining the word *robot* in his 1920 play, *R.U.R. (Rossum's Universal Robots)*. He also recognized one of the most salient points about human anatomy when he had one of the play's characters point out:

Well, anyone who's looked into anatomy will have seen at once that man is too complicated, and that a good engineer could make him more simply. So young Rossum [the inventor of the humanlike robots] began to overhaul anatomy and tried to see what could be left out or simplified.

Čapek was right. The biochemical reactions taking place within you at this moment are rarely, if ever, straightforward. For example, it takes more than twenty separate processes to accomplish something as vital as clotting blood when you cut yourself. It is a wonder that we haven't all bled to death long ago.

On the other hand, if the clotting process were too efficient, we might have all been wiped out by blood clots that cause heart attacks or strokes. There is a balance between too much and too little, and we cannot complain about a process that has helped keep our species from extinction.

The term *apoptosis* was introduced by Australian pathologist John F. R. Kerr, together with Scottish colleagues Andrew H. Wylie and Alistair R. Currie, in 1972. They also proposed that when apoptotic cell death goes wrong, it may play a role in many diseases, including Alzheimer's and cancer.

Researchers in the 1990s have found that most cells in the human body (perhaps all of them) manufacture a set of proteins that trigger the process of self-destruction. As long as the cell is operating normally and is performing its function in the body, the suicide mechanism will not come into play. But if the cell becomes infected or develops into a malignant tumor, the death-inflicting proteins are set into action.

Again, this is not a simple process. There are several possible triggers. The cells of the human body are not 100 trillion separate little entities. They influence one another constantly. Each cell is in contact with its neighbors and with distant cells, mainly through chemicals that serve as messengers. These chemical signals are complex and quite subtle. The cells react to them in ways that are not yet fully understood.

Healthy cells require chemicals called *growth factors*. An infected or malignant cell can stop using growth factors; this is the cellular equivalent of putting a loaded pistol to your head. When a cell stops producing growth factors, apoptotic death is usually the end result.

Alternatively, apoptotic death can be triggered if a cell receives external chemical messages that override the growth factors' influence. In other words, other cells in the body *command* the cell to commit suicide.

In an article in the December 1996 *Scientific American*, Richard C. Duke and colleagues David M. Ojcius and John Ding-E Young trace some of the ways a T lymphocyte can be induced to commit suicide. As we have seen, T cells are one of the so-called "white blood corpuscles" that patrol the bloodstream and lymph. Their task is to kill invading microbes and cells infected with viruses.

T cells are constantly being created in the bone marrow. To be successful killers, they must develop specialized receptors on their surfaces that can recognize not only invading cells and cells infected by viruses but the body's own cells as well. Otherwise, the killer T cells would attack the normal cells of the body. This is what happens in autoimmune diseases, such as arthritis and allergies. Immature T cells that do not develop these sensors properly undergo apoptosis. They kill themselves before they leave the bone marrow and enter the bloodstream.

Once in the bloodstream, T cells search for invading microbes or cells that have been infected by viruses. If its DNA is damaged (by a virus, for example), the T cell produces a protein called p53. Dubbed "the molecule of the year" in 1993, p53 was thought to be *the* self-destruct trigger for all cells. Additional research, however, showed that while p53 is present in most cases of apoptosis, it is not always required.

There is more than one trigger for cell suicide.

When a T cell circulating in the bloodstream discovers an invading bacterium, it binds to the invader, engulfs it, and literally tears it to pieces with oxidizing chemical enzymes that break up the microbe's molecular structure. The T cell also secretes proteins that cause inflammation, which is why a cut or abrasion of your skin reddens, swells slightly, and feels hot to the touch.

69

Once the invading microbes have been destroyed, the victorious T cells must commit suicide. If they do not, they will continue to cause inflammation, swelling, and fever and may even begin to attack the body's own cells.

There are at least two additional triggers for the suicide of cells that are no longer needed: one active and one passive.

The passive trigger is to deprive the cell of one of the chemical factors it needs to keep on living, a molecule called *interleukin-2*. Like a person deprived of water or food, the cell dies.

The active trigger involves a molecule called *Fas*, which sticks out from the T cell's surface like an antenna, while its other end penetrates through the membrane into the cell's interior. When a T cell enters chemical combat with an invading or infected cell, it starts to produce extra Fas, plus an associated molecule called *Fas ligand*. A few days after Fas ligand is first produced, the T cell kills itself.

The Apoptosis Program

While T cells and other lymphocytes are constantly being produced by the bone marrow and just as constantly killing themselves, most of the other cells in the body last much longer. Some of them, such as nerve and major muscle cells, must avoid death if the body is to continue functioning. If all of our cells contain suicide programs within themselves, at least some cells refuse to trigger that program easily.

Yet the suicide program is apparently a necessary part of our cellular machinery. When an ordinary cell—say, a cell in the lung—suffers damage to its DNA due to a genetic mutation or infection by a virus, if that cell is not eliminated, it could lead to cancer. The apoptosis program is there, buried deep inside the cell's workings, to get the cell to commit suicide before it begins to damage the cells around it.

Make no mistake. It's a dangerous world both outside our skins and inside them. Viruses can insert their own DNA into a cell and use the infected cell's machinery to make more copies of the virus, rather than carrying out the cell's usual task. The Epstein-Barr virus, for example, causes mononucleosis and can lead to cancer.

Cancerous cells, like the HeLa tumor cells, keep reproducing indefinitely. The Hayflick limit does not apply to them. That is why cancer kills. Unchecked, the tumor keeps growing, invading and crowding the body's organs, even breaking bones in some severe cases. Cancer cells have somehow "learned" how to evade apoptosis.

While scientists have unraveled some of the trigger mechanisms that lead to programmed cell death, it seems clear that these triggers must themselves be programmed by the cell's genes. After all, the genes are the command center for each cell. Everything the cell does is directed by the DNA of the genes.

Is there a "death gene" within each cell of our bodies, patiently waiting until a particular moment and then commanding the cell to commit suicide?

70

6

Youth and Death Genes

*Understanding how DNA works may be the greatest
scientific breakthrough in history.*

—LYNN MARGULIS

∞ IN NOVEMBER 1997 A TEAM OF JAPANESE RESEARCHERS
reported discovery of a gene that apparently suppresses
aging.

They named the gene *klotho*, after the Greek goddess
who spins the thread of life. In laboratory mice, when the
klotho gene is working properly, the mice live their normal
span of 2 to 3 years; when the gene is mutated so that it
cannot function properly, the mice die within 60 days.

The *klotho* gene also exists in humans.

The research team, led by Makoto Kuro-o, a physician
and molecular geneticist at the National Institute of Neu-
roscience in Tokyo, found that lab mice with defective
klotho genes quickly succumbed to a combination of arte-
riosclerosis, skin atrophy, osteoporosis, emphysema, and
infertility—afflictions that are associated with aging in
humans.

The *klotho* gene apparently codes for a protein that
circulates in the blood and somehow suppresses these
age-related ailments.

Scientists around the world were surprised by the Japanese finding. Although they are cautious in their acceptance of *klotho*'s relation to human aging, the discovery lends weight to the possibility that there are genes in human cells that can suppress aging. If they could remain active, we would not age.

Klotho may be a "youth gene." The *klotho* discovery may mean that there is a genetic basis for youthful vigor and resistance to the afflictions of old age. If so, the gene somehow stops functioning as we grow older. And once it stops functioning, we begin to suffer the ailments we associate with aging.

What about the other side of the coin?

Is there a "death gene"?

Deep within the cells of our bodies, is there a triggering mechanism that tells the cell when it is time to die? Is aging—senescence—truly inevitable?

Is there a built-in clock in each of our cells that sooner or later starts the process of aging and programmed cell death? If there is, can we prevent it from causing death? Can we, like the lowly paramecium, learn how to reset the clock of senescence?

In the third era of medicine, researchers began to look into the inner workings of the genes, down at the molecular level. They unraveled the structure of DNA and learned how this gigantic molecule uses a *genetic code* to control the operation and reproduction of the cell.

This new field of molecular biology is already showing promise of finding ways to combat genetic diseases: disorders such as diabetes, cystic fibrosis, and cancer that are caused not by infectious microbes or viruses but by defects in the genes.

The chromosomes are molecules of DNA. Strung along the chromosomes, like pearls on a necklace, are the *genes*—the molecular units that govern every aspect of the cell. Your genes determine the color of your eyes, the hormones that regulate your heartbeat, the production of red corpuscles in your blood, the uptake of calcium by your bones. Every physical aspect of your body is determined by the actions of the genes in your 100 trillion cells.

Do these genes also command your cells to grow old and die? Is senescence determined genetically?

Leonard Hayflick, who discovered that human cells have sharp limits to the number of times they will reproduce, is wary of the idea of a cellular clock that actually measures time in days or years. He agrees, however, that cells must contain mechanisms that count events such as cell divisions. Hayflick writes, "Most biogerontologists [biologists who study the processes of aging] agree not only that cells must contain multiple biological event counters but also that *these counters determine an organism's maximum potential life span.* [italics added]."

If cells contain counters ticking away like the meter in a taxicab, such counters must be controlled by the genes.

If we can understand how our genes work and map out which genes perform which functions in human cells, we may be able not only to correct genetic disorders but also to discover if death genes exist—and how to cancel their action.

The Master Blueprints

Genes are the master blueprints of our cells. While the mitochondria have their own small set of genes, it is the genes within the nucleus of each of our cells that hold the information that directs the construction, operation, and reproduction of the cells. When our genes are performing correctly, our cells work as they should. When there is a defect in one or more genes, we become susceptible to genetic disorders.

Death itself may be inevitable (so far), not because we

are destined to succumb to infectious diseases or genetic defects but because there is a suicide program deliberately written into our genes, like a hidden computer virus that eventually destroys all the computer's programs.

Alternatively, death may be the result of aging—we simply wear out, in time, like a poorly maintained automobile. We succumb to the accumulated damage caused by free radicals and glucose browning. Yet an automobile may last indefinitely if it is properly maintained. If tender loving care can extend a machine's life span, what about our own?

In the early years of our lives, our cells are capable of repairing damage inflicted upon them. As the years accumulate, however, these maintenance programs slow down. Repair becomes more difficult and eventually ceases altogether. Can we discover why the repair programs lose their effectiveness? Can we learn how to keep them active and thereby avoid aging and death?

Aging itself (or senescence) may be a consequence of genetic instructions or malfunctions. Either way, death is intimately bound to our genes. Understand how the genes work, and we may find how to forestall or even eliminate both aging and death.

Chromosomes and Telomeres

Inside the nucleus of eukaryotic cells are the long strands or filaments called *chromosomes*. Human cells have forty-six chromosomes, except for the sex cells, which have half that number. The chromosomes contain DNA. And DNA makes up the cell's genes.

Bacteria contain chromosomal DNA, just as all cells do, but in the prokaryotic bacteria, the chromosomes float freely in the cell, rather than remaining inside a nucleus, and in most bacteria, the DNA consists of a single chromosomal strand, usually curled into a loop.

Algae, amebas, paramecia, and many other species that biologists lump together under the heading of *protists* are also one-celled organisms. But unlike the bacteria, protists are eukaryotic; their cells contain nuclei that house their DNA-containing chromosomes. Multicellular plants and animals, the so-called "higher" life-forms, are all eukaryotes.

Eukaryotic chromosomes are capped at their ends by *telomeres*, which prevent a chromosome from attaching itself to another chromosome or looping itself into a closed circle. The telomeres may play an important role in aging.

75

Telomere Shortening

When a cell divides, it begins by making a copy of each of its chromosomes. This is called the *S phase* of cellular division. Some ten to twenty hours later the two sets of chromosomes pull apart to form two separate nuclei. Then the cell itself divides in two, each cell with a fully functional nucleus containing a full complement of chromosomes. The nuclear division process is called *mitosis*.

During the S phase of the cycle, the telomeres at the ends of each chromosome are reproduced along with the rest of the chromosome. But except for the germ cells of the ovaries and testicles, as the cells continue to reproduce, the telomeres become shorter and shorter. As the cells age, the telomeres shorten a little with each round of cell division. It's as if the cell runs out of material to build the telomeres—yet the telomeres are made of DNA, just as the rest of the chromosomes are.

If and when the telomeres disappear altogether, the chromosome ends begin sticking together, which makes it difficult (if not impossible) for the genes to work correctly. When genes cannot perform their functions properly, the cell cannot perform its functions properly.

Telomere shortening, then, seems intimately involved

in aging. Perhaps it is one of the *causes* of aging. Tumor cells, which are effectively immortal, produce the enzyme *telomerase* that rebuilds their telomeres each time the cell divides. HeLa cells, which have been reproducing since at least 1951, still exhibit telomeres that are just as long as they were nearly half a century ago and produce telomerase, just as they did nearly fifty years ago.

Michael Fossel, professor of clinical medicine at Michigan State University, says quite clearly, "Telomeres [are] the clocks of aging." He and other researchers believe that telomere shortening is responsible for cellular aging and, eventually, cellular death. And as we have already seen, we die because our cells die.

Most biologists are wary of this telomeric argument. To them, the idea that telomere shortening is *the* cause of aging seems too simple, too pat.

Yet early in 1998 researchers at Geron Corporation, in Menlo Park, California, and the University of Texas Southwestern Medical Center reported that they had nearly doubled the life span of human cells by injecting them with a gene that allows the cells to rebuild their telomeres and thereby avoid the Hayflick limit.

"These cells have an indefinite life span," said Calvin Harley, one of the researchers and a vice president of Geron.

Whether the telomeric argument is right or wrong, senescence and death definitely appear to have a genetic basis. If we can understand how the genes work, we may discover what causes apoptosis and senescent death.

The Double Helix

The genes are made of DNA.

DNA is the fundamental molecule of life. At the heart of each and every living cell, be it the cells of bacteria, algae, mushrooms, yeast, whales, Sequoia trees, or humans, is a wondrous, gigantic molecule called *deoxyribonucleic acid* (DNA).[11] The chromosomes of eukaryotic cells such as our own are basically long coils of DNA wrapped around protein cores called *histones.* Bacterial DNA floats loosely in the cell, without a histone backbone.

The DNA molecule's structure is a beautifully simple double helix. DNA is built like a flexible ladder that has been twisted into a coil. You might think of the DNA molecule as a huge coiled spring.

Using words such as *gigantic, long,* and *huge* may seem utterly out of place when talking about molecules. After all, these things are so small they can only be seen with powerful electron microscopes. Yet inside the nucleus of every one of the 100 trillion cells in your body is a super-squeezed coil of DNA which, if stretched out its full length, would extend more than six feet! Think about that the next time you pack clothes into a suitcase.

The double helix of DNA directs each cell's activities. Like a master computer program, the DNA-based genes tell the cell when to reproduce itself by dividing into two and how to make the proteins that are the building blocks of our cells. And when to die. (For details, see Appendix B: DNA.)

Turning Off the Death Clock

Again, though, the complexity of the human body comes into play. Some genes have more than one effect; the proteins that their blueprints produce play multiple roles in the life—and death—of our cells.

For example, recent research has shown that genes

---

[11] Well, not exactly *all* cells have DNA. Some "primitive" bacteria contain only the simpler RNA.

linked to breast cancer may play a key role in repairing damaged DNA.

In 1994 and 1995 researchers found that mutations in the genes *BRCA1* and *BRCA2* are responsible for most inherited breast and ovarian cancers.[12] In 1997 several studies showed that the proteins these genes normally produce are among the enzymes that detect damage to the cells' DNA and repair that damage.

When these self-repair genes are mutated, however (by a virus, toxins, and so on), they can apparently cancel the cellular clock that limits the cell's ability to reproduce and allow the cell to reproduce endlessly. A tumor starts growing.

Certain proteins, therefore, appear to be able to stop or reset the cellular clock that sets the Hayflick limit. Those proteins are produced by genes that carry their molecular blueprints.

Some of the proteins manufactured in our cells are triggers for apoptosis, as we saw in the previous chapter. There are specific genes in our cells, then, that direct the production of proteins, such as p53, interleukin-2, Fas, and the other proteins that lead to cellular suicide.

And there are other genes that code for BRCA1 and other proteins that cancel the cellular clock of the Hayflick limit and allow the cell to reproduce endlessly.

To understand how these genes work and how to gain safe control of their functions so that we may avert programmed cellular death if and when we want to is *the* major task of the fourth era of medicine.

---

[12]Biologists use italics to identify a gene, and normal type to identify the protein that the gene codes for. Thus *BRCA1* is the gene; BRCA1 is the protein it produces.

7

The Genetic Code

I don't want to achieve immortality through my work. . . .
I want to achieve it through not dying.

∞ THE MANY THOUSANDS OF DIFFERENT PROTEINS THAT comprise our bodies are each constructed from just twenty different molecular building blocks called *amino acids*. Somehow the genes in the nuclei of our cells carry blueprints for bringing together amino acids in the proper sequence to form all the various proteins.

How do they do that?

Once Watson and Crick determined the double-helix structure of DNA, in 1953, it was the astrophysicist George Gamow who suggested, the following year, that the subunits of DNA, called *bases*, might form a kind of code that carries the information needed to create new proteins.

Gamow was right. Within fifteen years, the complete genetic code had been deciphered, and researchers understood the way that DNA directs the production of amino acids, which are put together in the ribosomes to form proteins.

The DNA molecule twists around like a long spiral staircase, but it is easier to picture how it works if you think of the DNA molecule as a zipper that has been coiled around and around. The "teeth" of the zipper are chemicals called *bases*.

When DNA copies itself (a process called *replication*), the zipper unzips along its full length. New "teeth" are built along the entire length of each half of the molecule to match the "teeth" that have been exposed by the unzipping. When replication is finished, there are two identical molecules where there once had been only one. This is the basis for all life on Earth: replication.

When DNA codes for a protein, the molecule unzips only partially, exposing only a portion of its "teeth," or bases. That set of bases is a gene.

Three DNA bases form a code unit, rather like the dots and dashes of Morse code. Each so-called "base triplet" codes for a certain amino acid. The triplets are called *codons*. Each codon can be thought of as a letter in the genetic code.

The amino acids are the "words" formed by the "letters" of the DNA triplet codons. The "words" of the amino acids are put together to form "sentences"—complete protein molecules. (See Appendix C: The Genetic Code for details.)

Mutations

A DNA molecule may faithfully direct the production of proteins for all the years of our lives without making a mistake. It may replicate itself time and again precisely, without fail.

Yet mistakes do happen. In linking up the tens of millions of bases ("teeth") within the DNA molecule, now and then something goes slightly awry. DNA can be attacked chemically by dangerously reactive free radicals. Or perhaps a stray bit of energy from a cosmic ray or the natural background radiation that surrounds us happens to damage part of the molecule. Or toxic substances within the body—produced by an infecting microbe or by nicotine or narcotic drugs—can damage the DNA.

If the damage to the DNA is not quickly detected and repaired, the genes will produce faulty proteins and the cell's functions will be altered, perhaps fatally. When the damage reaches a critical point, the damaged DNA may trigger its inbuilt suicide command and the cell undergoes apoptotic death. We see this as aging, slowing down, and—when enough of our cells have suffered such a fate—we die.

But we have a natural defense against DNA damage, an enzyme known as *DNA polymerase,* the chemical repair-man. DNA polymerase is capable of recognizing damaged sections of the unraveled DNA strand, snipping out the damaged part, and replacing it with the proper material.

Like a faithful watchman patroling his beat, molecules of DNA polymerase move up and down the DNA helix, looking for damage and repairing it whenever damage is found. DNA polymerase can check about ten bases per second. While this is slow compared to the speed with which other enzymes work, DNA polymerase trades speed for accuracy. It is the molecular quality control system that keeps the DNA in the nuclei of our cells functioning properly.

However, no system is entirely perfect, and over time damaged sections of DNA are not repaired.

Such alterations in our DNA cause alterations in the proteins that the DNA manufactures. And when the DNA replicates itself and reproduces that alteration in the new molecule, we have what biologists call a *mutation.* The original DNA has been changed permanently, and the new version will faithfully reproduce itself, mutation included.

Most mutations are negligible. They make no effective difference in the way the cell operates, nor in the way the organism behaves. Some mutations are actually beneficial: They improve the organism's ability to survive in its environment. Although beneficial mutations are extremely rare, they have led to the diversity of life. Remember, you and I are the result of billions of years of such mutations. We are essentially the botched attempt of bacteria to replicate themselves.

Many mutations are harmful. They cause changes in the operation of the DNA that actually damage the cell's ability to continue functioning and that impair the organism's ability to survive.

Moreover, mutations are passed from parent to offspring. They can accumulate over many generations. Mutations that are harmless in a parent can become deadly in a child, when the child has acquired mutated DNA from both its parents.

Genetic Disorders and Gene Therapy

Once scientists realized that diseases such as diabetes, cystic fibrosis, Down syndrome, Parkinson's disease, Alzheimer's, and cancer are caused by genetic disorders, they began to search for ways to repair or replace malfunctioning genes.

The field of *gene therapy*, the central feature of the third era of medicine, is very new. As this is written, barely two thousand patients have received gene therapy, and none of them has been cured of the genetic disorder that afflicts her or him. Yet the first gene therapy trials in human patients began only in the late 1980s, and the concept holds enormous promise.

To make gene therapy successful, biomedical scientists must solve a number of problems. Like detectives, they must find out who the "criminal" is, how to catch him, and how either to rehabilitate the criminal or replace him with a law-abiding citizen.

What actually goes wrong among the genes when a person comes down with cancer, for example? Which of the 100,000 or more genes in the human body has gone wrong, turned criminal? Biomedical scientists have identified the genes responsible for many disorders and are doggedly on the trail of still others—including the possible gene or set of genes that triggers programmed cell death, apoptosis.

The biggest problem today is how to replace the defective genes with "good" ones. Researchers have developed several techniques, all of them experimental at present, none of them completely satisfactory.

In most cases, cells are removed from the patient and the therapeutic "good" genes are inserted into them in the laboratory. Then the corrected cells are replaced into the patient. This is the *ex vivo* (out of the body) method. The *in vivo* method inserts the genes directly into the patient's body.

The most common technique for inserting genes into the body is to use a viral *vector*. That is, a virus is harnessed to carry the new genes into the patient's cells.

The virus is genetically engineered to accomplish its mission. Molecular biologists alter the virus so that its disease-causing ability is disabled, and the desired therapeutic gene is inserted into the virus. Then the virus is put into the body, carrying its load of "good" genes to the defective cells like a Trojan horse (see Viruses, page 84).

There are other techniques for inserting therapeutic genes into the body. In one, fatty molecules called *liposomes* are used to carry the genes, rather than viruses. In another, "naked" DNA is inserted into the cells, without a carrier.

None of these techniques works perfectly. What researchers want is a sort of molecular homing pigeon that will go directly to the afflicted cells and replace the defective genes with the correct genes. Today's techniques are only a beginning toward that goal.

But they are a beginning that can eventually lead to the knowledge that will allow bioscientists to extend the human life span indefinitely.

VIRUSES

Viruses exist on the borderline between life and non-life. Much smaller than cells, only a few nanometers long (a few hundred millionths of an inch), they are little more than packages of DNA with a protein coating around them. Viruses can remain dormant for years, decades, perhaps millennia or longer. They can be frozen in liquid nitrogen or kept in airless vacuum chambers. They seem like inert specks of dust—until they come in contact with a living cell. Then they spring to life, puncture the cell's membrane, insert their own DNA into the cell's nucleus, and co-opt the cell's machinery into producing more of themselves.

In effect, viruses execute a "hostile takeover" of the cell and force the cell to stop making what the cell's DNA would normally produce in order to make more viral DNA. Eventually the cell bursts, flinging the newly produced viruses into the surrounding tissues, ready to take over more cells.

Second-era medicine had no defense against viral infections. Antibiotics work at the cellular level, not the molecular level of DNA. The body's own defenses against viral infections consist mainly of T lymphocytes that attack and kill the cells infected by the virus. This is somewhat like dynamiting the factory that has been taken over by a hostile corporation.

Third-era medicine offers some hope of learning how to disable or demolish altogether the viral DNA.

In the meantime, bioscientists are using genetically-engineered viruses as *vectors* to carry altered genes into the cells afflicted by genetic disease.

8

Gene Therapy

The impossible we do immediately; the miraculous takes a little longer.

—ATTRIBUTED TO U.S. MILITARY
CONSTRUCTION UNITS, WORLD WAR II

∞ ONCE THE GENETIC CODE WAS DECIPHERED, RESEARCHERS and physicians realized that it should be possible to deliberately alter or replace damaged genes. Thus the hallmark of medicine's third era has become *gene therapy*.

Gene therapy is in its infancy today, yet once molecular biologists understand the genetic basis of aging, gene therapy will become one of the major tools in extending human life span toward infinity.

Several genetic diseases are the targets for today's gene therapy trials, including diabetes, cystic fibrosis, and cancer.

Diabetes

Diabetes is one of the most prevalent genetic disorders. More than fifty thousand Americans die of diabetes each year.

The disease is caused by a person's inability to produce the hormone *insulin*. Normally insulin is manufactured in the pancreas, but in at least one type of diabetes—called *Type 1* or *juvenile-onset diabetes*—the beta cells of the pancreas cannot secrete insulin. This can be caused by inheriting a defect in the gene that codes for insulin production.

In some cases, the cause is not the gene for insulin itself but a defect in the segment of the pancreatic cells' DNA that regulates the rate at which messenger RNA can transcribe the DNA's blueprint for insulin. In other words, the pancreas can still produce insulin but does not produce it in sufficient quantity to meet the body's needs, because the cellular regulator that controls the amount of insulin produced has gone awry. Imagine a factory's assembly line slowing down because the factory's director is sleeping on the job, instead of making certain that production quotas are maintained.

Type 2 diabetes is called *adult-onset diabetes*, because it generally does not begin to show its effects until the patient is at least 20 years old. Researchers have associated the gene *obese* with Type 2 diabetes in laboratory mice. *Obese* codes for a protein named *leptin*, which is made of 176 amino acids. When injected into overweight lab mice, it causes weight loss.

In 1994 a team from the Howard Hughes Medical Institute and Rockefeller University, led by Yiying Zhang, cloned the *obese* mouse gene and determined that it plays an important role in determining the amount of fat the body will hold. News stories proclaimed that scientists had found "the" gene that controls body weight: an overstatement at best. The true importance of understanding the working of the *obese* gene lies in the promise of learning how to eliminate adult-onset diabetes.

Sickle-cell Anemia

You would think that defective genes, which lead to disease and death, would ultimately disappear from the gene pool. If the people carrying the gene do not live as long or as healthy lives as those who do not carry the defective gene, they should have fewer offspring and eventually die off. After all, how many Edsels are still on the road?

Yet nature is not always that simple. Some "defective" genes offer positive benefits. Sickle-cell anemia, a disease that predominantly afflicts Africans and African Americans, is a case in point.

Remember, a gene's job is to direct the cell's production of a certain protein. We have two sets of genes in our diploid cells. Usually, when one gene is defective, its healthy mate will take over the task of producing the necessary protein. It is usually when the cell inherits defective genes from both its parents that a genetic disorder manifests itself. Neither gene is capable of producing the needed protein.

87

A person carrying *one* sickle-cell gene does not come down with the disease. Instead, that "defective" gene produces a protein that helps the immune system to protect against the infectious disease of malaria. A person bearing that "defective" gene is protected against malaria, which is an endemic disease in the tropics. For people living in areas where malaria is prevalent, this "bad" gene is a blessing.

However, if the person inherits the sickle-cell gene from both parents, then he or she does indeed contract sickle-cell anemia: The victim's red blood corpuscles are deformed and unable to carry oxygen properly to the cells of the body.

Cystic Fibrosis

Once the "criminal" has been identified, once the defective gene is found, the problem then is to find a way to replace it with a gene that functions properly. This is where research stands today in regard to cystic fibrosis.

Cystic fibrosis is the most common serious genetic disease that affects Caucasians. It strikes one of every two thousand white children in the United States. Patients afflicted with cystic fibrosis suffer repeated lung infections and have difficulty breathing because mucus builds up in the lungs. In some cases, mucus also collects in the pancreas, intestines, and sperm ducts, interfering with digestive processes and even leading to male sterility.

Second-era medicine had no effective treatment for cystic fibrosis, except for a few experimental mucus-dissolving drugs. Most patients depend on daily physical therapy to dislodge lung mucus and doses of antibiotics to prevent infections from starting in the mucus-clogged organs. Still, few patients survive past age 30.

Cystic fibrosis may yield to the new third-era technique of gene therapy. Tests are already underway.

In 1989 researchers identified the gene that, when damaged, causes cystic fibrosis: *cystic fibrosis transmembrane conductance regulator* (*CFTR*). This gene normally codes for the production of a protein that allows cells to excrete salty chloride ions. For years, scientists had suspected that cystic fibrosis was linked to the body's regulation of salt, because people with the disorder have abnormally salty sweat. When mutated, instead of producing the protein that helps the cell get rid of chloride ions, the altered *CFTR* gene codes for a protein that helps to cause mucus deposition.

In the new gene therapy experiments, researchers are attempting to correct this defect by inserting properly functioning *CFTR* genes into cystic fibrosis patients. Three research teams from the National Heart, Lung, and Blood Institute, the University of Michigan Medical Center, and the University of Iowa College of Medicine are using similar approaches.

The scientists use a genetically engineered species of

virus as the delivery system for placing functional *CFTR* genes into the patient's cells.

The researchers fighting cystic fibrosis have enlisted a type of virus called an *adenovirus* as a vector. Adenoviruses can cause the common cold, so first the researchers disable the virus's ability to multiply once inside the human body. Then they place correctly working *CFTR* genes into the adenovirus. Adenoviruses have a specific affinity for the epithelial cells lining air passages in the nose and lungs; that is where they cause the miseries of a cold. It is also just where the researchers want to deliver the functional *CFTR* genes. The *CFTR*-bearing virus is administered to the patients through a nasal spray.

The most important goal of these tests is to see if the technique of delivering working genes through a viral vector can work—and work safely. Can the virus truly be disabled so that it will not reproduce once inside the human body but only deliver the functional *CFTR* genes?

More importantly, will the body accept the new genes and use them in place of the malfunctioning genes to produce the protein that the cells need? That is the key question. Also, if all goes well, how long will the beneficial effect last? Weeks? Months? Will the functional *CFTR* gene become permanently integrated into the body's cells, thereby curing the patient of cystic fibrosis?

One of the leading researchers in this field, Ronald G. Crystal of the National Heart, Lung, and Blood Institute, believes that the benefits from the therapy will last no longer than a few months, because the adenoviruses will not insert *CFTR* genes permanently into the chromosomal DNA of the cells of the patients' airways.

Yet there is another approach that may be much more helpful. In March 1997 a research team in New Orleans announced that they had cured laboratory mice that had a mutant gene similar to the one that causes cystic fibrosis in humans.

They injected a viral vector bearing the correct gene into mouse fetuses, through the mouse mother, in a procedure somewhat like amniocentesis. All thirteen of the baby mice were born with the correct gene instead of the defective one.

"Cystic fibrosis is a preventable disease," said J. Craig Cohen. In essence, the New Orleans researchers believe that cystic fibrosis is a birth defect that should be treatable—and cured—*in utero*. As of this writing, most scientists are wary of accepting such a bold claim on such thin evidence. But the evidence does seem striking.

Down Syndrome

Sadly, many genetic disorders are still beyond the reach of modern biomedical science. Even when researchers understand what causes the disorder, they have not found a way to correct it. Down syndrome is a case in point.

Normally, human cells bear forty-six chromosomes, half from each parent. However, *gametes* (egg or sperm cells) carry only half that number.[13] The normal sperm cell carries one copy of each chromosome, and the egg cell carries one copy. When sperm fertilizes egg—*voilà!*—the resulting zygote bears forty-six chromosomes. If all goes well.

Down syndrome is caused by a mutation in which the fertilized egg bears three copies of *chromosome 21*, rather

---

[13]Gametes are also called *germ-line cells*, although this can be confused with the idea of bacteria and other microbes as "germs." When used to describe sperm or egg cells, *germ-line cells* mean cells associated with sex and reproduction and have nothing to do with infectious microbes. Biologists speak of germ-line cells (or gametes) as distinct from *somatic cells*, which are the cells of the rest of the body. Somatic cells age. Germ-line cells do not.

than the normal two. What has happened is that a mistake was made when the sperm or egg cell divided from the precursor cell that produced it. Other cells undergo mitosis when they divide; the chromosomes duplicate their DNA and then the nucleus divides into two. Sperm and egg cells undergo *meiosis* (from the Greek word for "diminishing"). During meiosis, the gametes take only one copy of each chromosome. Thus, while all the other cells of your body are diploidal, your gametes (or germ-line cells) are haploidal; each sperm or egg cell bears only one copy of each chromosome.

But in persons carrying Down syndrome, the germ-line cell takes both copies of chromosome 21 from its precursor cell. When this gamete becomes a fertilized egg, it is therefore carrying three copies of chromosome 21 instead of the normal two. Chromosome 21 happens to be the smallest of all the human chromosomes, containing "only" some 50 million base units; *chromosome 1*, the largest, contains almost 300 million bases.

In most cases, this genetic mistake is so drastic that the fetus aborts spontaneously; the mother suffers a miscarriage, usually without realizing that the fetus was afflicted with Down syndrome. Babies born with Down syndrome are usually mentally and physically retarded, often severely. Today Down syndrome is responsible for about one-third of all the cases of mental retardation in industrialized nations. Patients who live to the age of 50 almost inevitably come down with Alzheimer's disease as well.

Until recently, few Down patients lived much beyond 20. Many died from simple neglect; they were regarded as freaks, to be hidden away and forgotten. Even the very definition of a Down syndrome patient, as recently as the 1970 edition of the Encyclopaedia Britannica, used the word *monster*. Thanks to improvements in health care and understanding of the needs of the retarded, Down patients are living longer and are regarded as monstrous only by the ignorant.

I have a friend who is a Down syndrome victim. He is nearing 30, although his mental abilities are more like those of a child. Thanks to the loving care of his father and stepmother, he is living in his own apartment, works in a fast-food restaurant, and has a satisfying social life.

In 95 percent of Down cases, the gamete bearing the extra chromosome 21 comes from the mother. Women in their early teens and women near menopause are more likely to bear Down syndrome babies. Older women are more at risk than any other age group, "strengthening the ovary's dubious honor of being the most rapidly aging organ in the body (except for the placenta)," in the words of British biologist Roger Gosden. Women aged 40 or older have a 1 in 100 chance of having a Down syndrome child, while in younger women the chances are 1 in 1,000 or less. Except for early teenagers.

Down syndrome can be detected by prenatal test, although many parents choose not to be tested because then they would face the agonizing decision of whether or not to abort the pregnancy. To this date, there is no known way to cure Down syndrome. When the genetic defect is discovered in a fetus, it is already too late to do anything but suggest abortion. It should be possible to detect the extra copy of chromosome 21 in a woman's unfertilized eggs or a man's sperm cells, but such genetic screening tests are not yet standard procedure— and may never be.

We will examine in a later chapter the reactions that have arisen to prenatal testing and genetic screening programs that can detect genetic faults in adults or unborn babies. The psychological, social, and economic impacts of such screening are as important as the moral and ethical implications.

CARDIOVASCULAR DISEASE

Not all medical successes are due to scientific breakthroughs.

Cardiovascular disease is the leading cause of death in the United States, killing nearly a million Americans per year, yet the death rate from this set of ailments dropped 34 percent between 1980 and 1990.

Diseases affecting the heart and blood circulatory system are not necessarily genetic, although a genetic predisposition is one of the major criteria for atherosclerosis, in which fatty deposits in the arteries block blood flow and can lead to *ischemic* heart attacks.

The recent reduction in cardiovascular death rate has been attributed to two sources:

1. improved surgical techniques (such as coronary artery bypass procedures) and better medications for high blood pressure

2. changes in lifestyle (such as reducing intake of high-cholesterol foods and stopping tobacco smoking)

These lifestyle changes reflect the impact of a broad effort at informing the public about the advantages of a low-fat, low-cholesterol diet, the dangers of tobacco, and the benefits of exercise.

Some people are predisposed genetically to coronary artery disease. Their arteries' inner walls are grooved, rather than smooth, and offer better sites for holding the fatty deposits that turn into artery-clogging plaque. Genetic research offers the possibilities of identifying persons at risk so that they can take preventive dietary and lifestyle measures early in life.

Eventually, it may become possible to alter the gene or genes that cause the condition, thereby reducing people's risk of accumulating such deposits in their arteries.

Cancer

Of all the genetic defects that assail the human race, cancer is the most prevalent, the most dangerous, and the most dreaded. More than 500,000 Americans die of cancer each year; it is the second-leading cause of death in the United States. (Cardiovascular disease is first. See Cardiovascular Disease, page 93.)

Almost worse than those fatality numbers is the fear stirred by the pain and suffering cancer can cause. Cancer is insidious; its causes are partly genetic, partly environmental. It seems to strike almost capriciously. A person's family can be riddled by cancer, yet a particular individual never comes down with the disease. Alternatively, even if there is no history of it in a person's family, a particular individual can contract cancer. Others can live or work in environments that are known to be carcinogenic and never be affected.

Yet it is an important link in the quest for immortality, because cancer cells have learned how to evade the limits of aging; they have become immortal.

Millions of people suffer from various types of cancer, and each of us lives in its shadow. Cancer is so difficult to deal with because it comes in so many different forms and because it seems to be triggered by both genetic predisposition and environmental factors (see Types of Cancer, opposite). In essence, cancer is both a genetic and infectious disease, although the "infection" does not necessarily come from an invading microbe.

Cancer cells are riddled with genetic defects: broken chromosomes or the wrong number of chromosomes. In a normal cell, such defects would be devastating; the cell would be unable to produce the proteins it needs to go on living. It would be unable to function. But a cancer cell has no function, as far as the rest of the body is concerned. Its only "function" is to make more copies of it-

TYPES OF CANCER

There are many types of cancer:
- *carcinoma*: a general term for cancer of any epithelial tissue, such as glands and the linings of the lungs and digestive organs
- *melanoma*: cancer of the skin's pigment cells
- *leukemia*: in which the white blood cells reproduce uncontrollably. Leukemia in mice and chickens is caused by a virus.
- *lymphoma*: cancer of lymphatic tissues
- *myeloma*: cancer of the plasma cells, which originate in bone marrow
- *sarcoma*: cancer of connective tissues, such as muscle or bone. Sarcomas can also affect organs, such as the bladder, kidneys, liver, lungs, and spleen.

self, to keep on multiplying, to keep the resulting tumor growing.

How does a normal human cell turn cancerous? It could be altered by an invading virus. More often the cause is an environmental effect, such as free radicals, toxic chemicals, or ionizing radiation. These effects damage the genes. They are called *acquired* or *induced* mutations. That is, the patient was not born with the genetic defect; it was acquired by the individual's exposure to a carcinogenic agent.

"Cancer is . . . a disease of the DNA itself," says George Klein, chief of the tumor biology department of the Karolinska Institute of Stockholm.

In cancer, the cell's nuclear DNA is altered either by a virus or by environmental damage from toxins or radiation: acquired mutations. The damaged DNA can no longer regulate the cell's growth, and a tumor begins.

Some viruses insert their own DNA into the cell's chromosomes, and the cell begins to multiply endlessly. Fortunately, the immune system usually recognizes such altered cells and kills them.

Other viruses are subtler. They have no DNA of their own; their genetic information is carried on molecules of RNA. But when they invade a cell they copy their own RNA into double-stranded DNA and then splice their DNA into the cell's DNA. Viruses, so "simple" that they exist on the borderline between living and nonliving, can do the kind of gene-splicing that biologists only learned to accomplish some twenty years ago!

There is no single gene that can be said to cause cancer. Rather, human cells contain several dozen genes known as *proto-oncogenes* (potential cancer-causing genes). The proto-oncogenes are normally involved in regulating cell growth. For the cell to turn cancerous, one or more of the proto-oncogenes mutates into an *oncogene*, a gene that orders the cell to reproduce itself endlessly.

Another group of genes involved in cancer is the *tumor suppressor genes*, which make proteins such as p53 that control cell division. Normal cells make very little p53, but when a cell is damaged (by toxic chemicals, radiation, oxygen starvation, and so on), p53 levels rise swiftly and prevent the cell from reproducing or even trigger programmed cell death, apoptosis. In most tumors, however, p53 is mutated and cannot do its job. The cells multiply uncontrollably.

Some genes involved in DNA repair can be mutated and become oncogenic. These genes control the production of DNA polymerase, which constantly checks the cell's DNA, searching for mistakes whenever the DNA replicates itself and repairing any defects so that the genes are copied properly. When these *repair genes* are mutated, DNA replication goes haywire, leading to errors in the copying that can cause still more changes—mutations—in the copied genes.

Through these mutations, cancerous cells lose or destroy the inner mechanisms that regulate cell reproduction. They throw away the cellular clock that sets the Hayflick limit and become essentially immortal. Tumors grow and continue to grow, invading and crowding out the healthy organs of the body, taking up more and more of the body's blood supply of nutrients and oxygen until—if unchecked—they eventually kill their host.

No single mutation changes a cell's genes so drastically that it becomes cancerous. It takes a series of genetic changes to destroy a cell's growth control systems, a number of genetic "hits." For example, there are two groups of proto-oncogenes in the cell, and some members of both groups must suffer mutations and become oncogenes if the cell is to become cancerous. Besides that, we do have some natural, built-in safety factors at work within our cells; otherwise, cancer would be much more prevalent than it is.

Actually, the human body protects itself against cancer much better than smaller animals do. Steven N. Austad points out that human cells are apparently 100,000 times more resistant to cancer than mouse cells. He comes to this conclusion by estimating that humans weigh about 3,300 times more than mice, which means humans must have roughly 3,300 times as many cells, since the cells in both mice and men are about the same size. Humans live about 30 times longer than mice (70 or 80 years vs. 2 or 3). Multiply these factors, 3,300 x 30 = 99,000. Round it out to 100,000 for the sake of argument.

But wait, there's more. Elephants contain roughly 40 times the cells we do; whales, 600 times. Yet these very large animals live about as long as humans, so they must be even more resistant to lethal cancers than humans.

Instead of studying mice in our cancer laboratories,

Austad hints, we should be studying elephants and whales! The practical problems seem intractable, however. Besides, the short life spans of rodents mean that researchers can study thirty-some generations of mice or rats in the time it would take a single elephant to go through its life.

Human cells actually do have genes that defend against the effects of the mutant oncogenes. These are called *antioncogenes*. Cancer can only start if these antioncogenes are either destroyed or mutated to such a degree that they lose their defensive effectiveness.

The picture is never quite that simple, however. There are so many different types of cancer, and the turmoil raging in the cell's genes and chromosomes is so destructive, that it becomes extremely difficult to pin down exactly what has gone wrong in the cell. Biologist Christopher Wills compares the situation to that of investigators puzzling over a plane crash, trying to piece together what went wrong from the twisted bits of wreckage strewn along the ground.

The irony of ironies is that cancerous cells have found the trick of immortality. They are not subject to the Hayflick limit. They continue to reproduce themselves indefinitely. The well-known HeLa cells have been chugging along since 1951 and show no signs of slowing down.

Not all tumors are dangerous. "Benign" tumors (or *neoplasms*) do not undergo *metastasis*; that is, they do not spread to other parts of the body. "Malignant" tumors do metastasize. This is why early detection of a malignancy is so important. The tumor must be destroyed or surgically removed before it begins to metastasize.

Melanomas, cancer of the skin's pigment cells, are one of the most dangerous forms of cancer because they quickly undergo metastasis. Melanomas are often confused with *basal cell carcinomas*, which are the most common form of skin cancer. Perhaps the least dangerous form of cancer because they are easily detected and rarely metastasize, they can still be lethal to persons who spend a lot of time in the sun; basal cell carcinomas should be treated by a dermatologist.

Gene Therapy for Cancer

In all too many cases, the second-era therapies for cancer are almost as bad as the disease itself. Surgery, to be effective, should be performed early enough so that the tumor is still localized and can be removed in its entirety. Even then, there is no guarantee that all the cancerous cells have been excised.

Chemotherapy and radiation have enormous side effects, ranging from bouts of nausea to hair loss to damage to otherwise healthy organs. In effect, most forms of chemotherapy and radiation therapy are like using shotguns to hit a small target. The target may be obliterated, but a lot of territory near the target can be damaged, too.

Several *oncogenic* (cancer-causing) viruses have been identified, and a cancer vaccine has even been developed—for chickens. Yet it seems clear that viral infections and/or damage to the cell stemming from environmental causes wreak their destruction at the level of the genes.

Several types of gene therapy are being tested against cancer. One technique is to insert genetically-altered cells into the tumorous tissue to produce toxic molecules that will kill the cancer cells. Other techniques are aimed at correcting the genetic mutations that lead to the formation of oncogenes or to stimulate the body's natural production of healthy, active antioncogenes. Klein suggests using powerful *growth suppressor genes* or genes that promote apoptosis.

One of the most promising areas under investigation is the effort to understand how cancerous tumors evade the body's natural immune system. Most tumors are quickly detected and destroyed by the immune system's lymphocytes. Those that grow into threatening cancers, however, have "learned" how to camouflage themselves chemically so that they appear to the lymphocytes to be normal cells.

The body's cells produce chemical "signatures" that the lymphocytes recognize. It is as if the police officers patroling a neighborhood recognize the area's law-abiding citizens. When a stranger shows up, a cell that does not produce the proper chemical signature, the lymphocytes attack and destroy it. Some forms of cancerous cells are able to produce the correct chemical signal, so the lymphocytes do not attack them. In effect, the stranger disguises himself as one of the local citizens to evade detection by the cop on the beat.

Bone Marrow Transplants

The various tests of gene therapy for cancer (and other ailments) are so new that they have not proven themselves and have years to go before they enter standard medical practice.

"Most of these approaches have yet to pass even the most preliminary clinical tests," says R. Michael Blaese of the National Institutes of Health.

One use of gene therapy that *is* clinically useful in cancer treatment is in the area of bone marrow transplants.

The bone marrow produces white and red blood corpuscles. In leukemia, the white cells reproduce uncontrollably and attack the body's normal cells. At least one form of leukemia is known to be caused by a defective chromosome, called the *Philadelphia chromosome*, in which the genes have been mutated by environmental effects, such as ionizing radiation.

The disease can often be arrested by transplanting bone marrow from a healthy person into the leukemia patient after first bombarding the patient with very high doses of radiation to kill any cancerous cells in the patient's own marrow. The patient's marrow is so devastated by the radiation therapy that the patient would quickly die if a bone marrow transplant were not available.

While this technique should eliminate the leukemia, the disease sometimes reappears. Either the radiation failed to kill all the cancer cells in the patient or there were some undetected cancer cells lurking in the apparently healthy marrow that was transplanted.

Researchers needed a way to trace the transplanted cells once they were inserted into the patient. To do this, they insert a stretch of DNA from a bacterium into the marrow cells that are to be transplanted. The harmless bacterial DNA serves as a unique "flag" by which the scientists can trace the presence of the inserted marrow cells. If the bacterial "flag" shows up in the patient's blood, it confirms that the new marrow is performing as it should. If the leukemia returns and the cancerous cells carry the bacterial marker, it proves that the inserted marrow bore undetected cancer cells.

Thanks to this technique, scientists have discovered that in certain types of cancer it may be necessary to treat the new marrow itself before transplanting it into the patient.

Even that does not always work, unfortunately. The world-famous planetary astronomer and author Carl Sagan died of *myelodysplasia*, a disease affecting the stem cells that produce red and white corpuscles, despite half a dozen marrow transplants.

However, progress against cancer is being made. This

in itself will inevitably lengthen average life expectancy by reducing or removing one of the major killers. The knowledge and experience in gene therapy gained by researchers will also lead directly to extending life span toward immortality.

AIDS

While AIDS—*acquired immune deficiency syndrome*—is not a genetic disease, roughly 10 percent of all the gene therapy research currently underway is devoted to combating infection by the human immunodeficiency virus: HIV. (By far the greatest proportion of gene therapy research, nearly 50 percent, is focused on cancer.)

HIV attacks the white blood cells, the lymphocytes that are the guardian warriors in the body's immune system. When the lymphocytes are so devastated by HIV assault, the immune system breaks down and the victim becomes susceptible to any number of infectious diseases. The result is invariably fatal—unless the HIV invasion can be curbed.

As in the work on leukemia, gene therapy research on AIDS is aimed at the stem cells of the bone marrow, which produce the immune system's lymphocytes. The hope is to be able to find genes that resist HIV attack and then insert them into the lymphocytes.

One effort is attempting to use altered HIV-infected cells as the carrier (or vector) for bringing resistant genes to the immune system's cells. The reasoning is that the HIV cells are attracted to the immune system cells; that is what causes AIDS in the first place. If HIV cells can be disabled, their disease-causing genes removed and therapeutic genes inserted in their place, then the altered HIV cells may become the ultimate biomedical Trojan horse and the ravages of AIDS may be overcome.

Saving the Gene Pool

Gene therapy will become increasingly important as time goes by, not only as a means of helping individuals with genetic diseases but as the way to keep our gene pool from being swamped by genetic defects.

Very often, the answer to yesterday's problem becomes the problem of today. In earlier generations, individuals who bore genetic diseases died early in life, so young that they left few, if any, offspring. Genetic disorders eventually died with them. The pitiless forces of nature drove the genetically disadvantaged into extinction—and their genes with them.

More recently, however, modern medicine and health care (and changes in attitudes toward the disadvantaged) have allowed persons with genetic diseases to live longer and have children—who carry the same deadly genes. Thus the gene pool, the common heritage of our species, is acquiring genes that can cause more and more cases of Down syndrome, sickle-cell anemia, diabetes, and so on.

If this trend continues, eventually everyone will be in danger of inheriting such genes. Gene therapy, which offers the hope of repairing or replacing unhealthy genes, can correct this problem. We will need gene therapy, if for no other reason than to salvage the gene pool.

In a sense, we are the victims of our own success. By helping the genetically disadvantaged to survive, we endanger our common heritage. But gene therapy can change that and end the threat.

Genes for Aging

Genes that control aging have been found—in the cells of a microscopic worm and the humble and ubiquitous yeast.

S. Michael Jawinski, professor of biochemistry and molecular biology at Louisiana State Medical Center, has identified nine genes that influence the life span of certain species of yeast cells.

Yeast is a single-celled fungus that can convert sugar into alcohol and carbon dioxide. Yeast cells can be found almost anywhere, including the air we breathe. Various species of yeast have been used since time immemorial in baking, brewing, and making wine and liquors.

Two of the most important genes in regulating the yeast cells' life span are known as *RAS1* and *RAS2*. They help to govern the cell's interior balance of energy, the condition biologists call *homeostasis*. *RAS1* apparently shortens life span, while *RAS2* extends it. By manipulating the *RAS2* gene to produce more of its proteins than normal, the cell's life span is increased.

Researchers at Harvard University, studying a microscopic roundworm called *Caenorhabditis elegans*, found that mutations to the worm's *daf-2* gene can put the worm into a state of near-hibernation, slowing its metabolism and extending its life considerably. *C. elegans* is a nematode that lives in soil. Its normal life span is about 14 days. However, when faced with food scarcity or other threats, such as lack of water, tiny *C. elegans* can put itself into suspended animation for periods of 2 months or more.

The Harvard researchers located the gene responsible for this trick, *daf-2*, cloned it, and found that it is related to a gene in humans: the insulin receptor gene, which helps cells get rid of sugar in the blood.

As we will see, experiments indicate that a near-starvation diet (about 50 percent fewer calories than normally consumed) can lengthen life span by 25 to 40 percent in laboratory rodents, although the reasons are not yet clear. The Harvard roundworm work suggests that some aspects of aging can be attributed to insulin receptor activity, since the worms with mutated (and hence less active) *daf-2* genes live longer, as do the dieting mice, whose diets leave insulin receptors little to do.

Most likely, the insulin receptor gene is linked to the metabolism of glucose and, as we have seen, glucose browning is one of the principal processes of aging.

If this work with *C. elegans* and yeast cells can be extended to humans, the genetic basis for aging may be uncovered and human life spans lengthened by gene therapy.

It seems certain that the prospects for life extension and immortality depend crucially on our understanding of our genes: how they function and how their functioning might be deliberately controlled.

9

The Human Genome Project

Only once would I have the opportunity to let my scientific life encompass the path from [the] double helix to the three billion steps of the human genome.

—JAMES WATSON

∞ IF SCIENTISTS CAN EXTEND THE LIFE SPANS OF YEAST CELLS and microscopic worms, why can't they extend human life span? The simple answer is that they will, once they learn enough about the way human genes work. Yeast and roundworms are much easier to work with and understand, but the work with them will serve as the platform on which we reach for human immortality.

Genetic diseases and gene therapy are at the forefront of current biomedical research. Scientists have identified specific genes that are responsible for certain disorders and are devising ways to alter those genes or remove them entirely and replace them with properly functioning genes.

The techniques of gene therapy that are now being developed will undoubtedly be used when today's research identifies the genes that can modify or control the rate at which we age—and perhaps even reverse the aging process.

As we have seen, senescence and death itself may be genetically programmed, triggered by genes that slow and ultimately stop the cells' activities.

If we understood all the genes in the human chromosomes, if we had a map of which genes performed which functions, we would be able to find out exactly which genes are involved in cancer (the proto-oncogenes) and learn how to protect them—or, at least, repair them before they turn into active oncogenes.

We could pinpoint the genes that activate programmed cell death and find out how to avert cellular senescence.

We could discover how cancerous cells "learn" to evade the Hayflick limit and extend their lifetimes indefinitely. We could even learn how to reset the cellular clock and extend our life spans indefinitely.

To paraphrase Shakespeare, the key to immortality is not in the stars but in our genes.

We could accomplish all that if we had a complete map of all our genes. That is what the Human Genome Project is trying to determine.

Big Research vs. Small Research

The *human genome* is the sum total of all the genes in the forty-three human chromosomes: every one of the 3 billion bases in the chromosomes. The Human Genome Project is the effort to identify each and every one of our 100,000 genes and map out where along the chromosomes they are placed.

The Human Genome Project is the offspring of several researchers. In March 1986 the Italian Nobel laureate virologist Renato Dulbecco, then with the Salk Institute in California, suggested in an editorial in the prestigious journal *Science* that it should be possible—and certainly desirable—to map the entire human genome. This was the result of earlier discussions and conferences among bioscientists from many universities and government agencies. That same month of March 1986 saw the first Human Genome Conference in Santa Fe, New Mexico.

Two major government agencies are directing (and funding) the Human Genome Project. One is the National Institutes of Health (NIH), which funds most of the biomedical research conducted in American universities and hospitals. The other is the Department of Energy (DOE), which is vitally concerned with the health issues involved in nuclear power. The two agencies complement each other to a considerable extent. The NIH specializes in "small" research, studies undertaken by relatively small groups of university or hospital investigators. The DOE can handle "big" research, projects that involve massive installations, large staffs, and complex administration.

Both large and small research talents are needed for the Human Genome Project. On the one hand, probing the chromosomes to map out the genes takes the talents of many NIH-type scientists. On the other hand, the genome is so huge that coordinating and managing the task of mapping it out completely is more like the Apollo Moon shot project than a typical university study program.

In September 1988 James Watson, who had won the Nobel prize for his codiscovery (with Francis Crick) of the double-helix structure of DNA, agreed to head the Human Genome Project. He has since stepped down from that responsibility, but his presence in the project's early years lent the program the kind of prestige that convinced scientists all around the world to join in the effort. His presence also helped to convince decision-makers in Congress and the White House to fund the project adequately.

"A Buck a Base"

The goal of the project is to have a complete map of the human genome by the year 2005. The cost has often been pegged at $3 billion. There are some 3 billion bases in the human genome, which combine to form about 100,000 genes. The in-joke among genome researchers was that the $3 billion price tag amounted to $1.00 per base.

This cost is going down significantly as faster, more efficient ways to map the genome are developed. To date, the program has consistently progressed faster than the original projections; if current rates of progress are merely maintained, the entire human genome will be mapped before 2005 and at a cost far below $3 billion: considerably less than "a buck a base."

In 1991 the project received a budget of $60 million from the NIH and an additional $41 million from the DOE. That same year, AIDS research in the United States was funded at a total of $800 million, nearly eight times the level of the Human Genome Project. Compared to the federal budgets for the Department of Defense (more than $200 *billion* per year) and social welfare (more than $700 *billion* per year), the Human Genome Project is minuscule—yet it offers an incalculably huge payback.

Still, the project is the largest organized effort ever undertaken in biology, akin in scope and importance to the physicists' Manhattan Project, which produced the first atomic bombs.

What the Human Genome Project is attempting to do is to write a sort of encyclopedia that will contain all the information stored in the 3 billion bases of our chromosomes.

My copy of the fifteenth edition of the Encyclopaedia Britannica (1972)[14] has 54 characters per line, 79 lines per page, and 2 columns on each page. This adds up to 4,266 characters (a letter or a space) on each page. The average number of pages per volume is 1,149, so each volume of the encyclopedia averages 9,735,012 characters. Let us be

---

[14]I have the 1996 edition on CD-ROM.

generous and say there are 10 million characters per volume.

If each character in the encyclopedia, each individual letter or space between words, represented one base in the human genome, it would take 300 volumes (at 1,149 pages each) to contain all the information stored in the 3 billion bases of your chromosomes. If each volume is 3 inches thick, the complete encyclopedia would take 75 feet of shelf space!

That is how big the job of mapping the entire human genome is.

Researchers around the world have tackled this tremendous task, and unless some unanticipated hitch crops up, the goal of mapping the complete genome by the year 2005 seems well within reach. But even if it takes longer, the task is well worth doing, much more important to our health and well-being than mapping the face of the Moon or the bottom of the oceans. The important factor now is to keep the project going, to keep the level of funding adequate to the needs of the researchers.

Even if the project eventually costs $3 billion (and it will cost considerably less), that level of funding is little more than 1 percent of our *annual* outlay for the Department of Defense. On an annual basis, the American people spend a thousand times more each year for pizza, cosmetics, or pet food.

Results are already coming in and are being passed on to the research community in general. Investigators have found that the chromosomes have vast stretches of seemingly "nonsense" areas, nothing more than boring repeats of a few bases over and over again, thousands of times. These "desert" regions do not appear to code for proteins; their functions—if any—are unknown.

As active genes are located and their functions are determined, other researchers are already using the information to further their own investigations of genetic disorders and the gene therapies that may correct or alleviate them.

Yet biologist/author Christopher Wills has pointed out that "simply determining the sequence of all this DNA will not mean that we have learned everything that there is to know about human beings, any more than looking up the sequence of notes in a Beethoven sonata gives us the capacity to play it. In the future, the true virtuosos of the genome will be those who can put the information to work and who can appreciate the subtle interactions of genes with each other and with the environment."

Genotype vs. Phenotype

The truth is that we are not simply the mechanistic result of our genes. The human body is much more complex than that. Mapping every one of your genes will be enormously helpful for your health and well-being, but the map is not the territory and knowing that you bear a gene that can lead to diabetes (for example) does not mean that you will automatically contract the disease.

Biologists speak of the *genotype* and the *phenotype*. The sum total of all the genes you carry is your genotype, the blueprint of you. But as anyone who has been involved in the construction of a house knows, the blueprint and the finished building do not always agree exactly.

Your genes interact with one another and with the environment in which you exist to produce a unique individual: you, a person of a certain height, weight, eye color, quality of voice, sense of smell, and so on. That is your phenotype, the individual you, as opposed to the blueprint.

Deep within your cells lurk oncogenes, but that does not automatically mean that you will contract cancer. Biomedical researchers and ethicists alike warn against *genetic determinism,* the concept that your genes alone determine your fate. The situation is not that simple.

For example, a gene called *Apolipoprotein E* is linked to atherosclerosis, the buildup of cholesterol plaque in the arteries that often results in heart attacks. *ApoE* is involved in processing fats in the foods we eat. There are several different forms of this gene.[15] People with the ∈4 allele of *ApoE* have higher blood cholesterol and higher LDL (low-density lipids, the "bad" cholesterol) levels. Finland shows a very high frequency of the ∈4 allele and also the highest rate of atherosclerosis in the world. Japan has the lowest rate of atherosclerosis and also one of the world's lowest frequencies of ∈4. So ∈4 can be said to be the gene responsible for atherosclerosis, right?

Not entirely.

The highest known frequency of the ∈4 allele is in Papua New Guinea, where atherosclerosis was virtually unknown until very recently. Before Western civilization came to the region, Papuans ate very little fat: less than 5 percent of their diet was fats, compared to 30 to 40 percent in the typical American diet. The jungle-dwelling Papuans were not particularly health-conscious, they simply did not have a lot of fatty foods available to them. Also, they led very strenuous lives.

Low fat intake and lots of exercise equals low incidence of atherosclerosis, despite the genes. Papuans who have changed their lifestyle to a more "modern" mode,

---

[15]Most genes come in slightly different forms in different people. Geneticists call the different forms *alleles.* Think of alleles as the slightly different versions of a particular automobile: One has a six-cylinder engine instead of a four, another has a CD player with its radio instead of a tape player, and so on. They are still the same make and model of car, only the "options" are different.

including less exercise and more fatty foods, die of heart attacks at relatively early ages.

The genes are there, but they operate in the environment in which the person exists. You may carry the ∈4 allele of *ApoE* yourself and never be bothered by atherosclerosis, if your living habits do not encourage the deposition of fatty plaque on your artery walls. On the other hand, you don't need to bear that particular gene to have a heart attack. Your genes are not the be-all and end-all of your existence.

Still, if you do have the ∈4 allele, wouldn't it be helpful if it could be identified and then altered into a less threatening allele of *ApoE*?

While it is certainly true that a complete map of the human genome will be as helpful to biomedical progress as AAA road maps are to motorists traveling into territory that is new to them, genes are not static things; they are dynamic, they interact with each other and the environment.

A complete map of your individual genotype would certainly be of enormous help in determining the risks you run for contracting, for example, Parkinson's disease; but that does not mean you will inevitably come down with Parkinson's.

However—and this is the vital point—if a genetic screening can reliably identify a mutant gene in your genotype (again, let us say it is the gene for Parkinson's) and if a technique for altering or replacing that gene with a healthy one exists, then repairing or replacing that defective gene will reduce your risk of contracting Parkinson's to zero.

Whether the whole human genome is mapped by 2005 or some later date, whether the effort costs $1.00 per base or some fraction of that figure, eventually the entire human genome will be mapped. Biomedical scientists will learn at last exactly which genes code for which proteins.

We already know that certain proteins, such as p53, can trigger programmed cell death. Others, such as the insulin receptor gene, apparently influence glucose browning. Once the entire human genome is mapped, scientists will be able to identify all the genes involved in senescence and death.

And once identified, those genes can be altered to suit our needs or desires.

10

Aging and Death

Death devours all lovely things.
                    —EDNA ST. VINCENT MILLAY

∞ I WATCHED MY FATHER SLOWLY DIE OVER THE COURSE OF a quarter century.

He was an active, vigorous man who had worked at a hard, physically demanding job from the time he was a teenager until he retired at 65. In his earlier years, he had been a semiprofessional baseball player, a scrappy infielder whose idols were the Philadelphia Athletics championship teams of 1929–31. Later he got into bowling; it was his favorite hobby.

At 65 he was very much the way I remembered him from twenty years earlier. His hair had silvered and thinned; some of his teeth were not his original equipment. He needed reading glasses. But he was still physically fit and mentally alert.

Gradually, over the years, his vigor faded. He had a minor stroke when he was in his seventies, and his physical deterioration accelerated after that. By the time he was 80, he was largely housebound. He lived with my sister and her family. I saw him occasionally, and each time he seemed a bit more diminished, weaker, less alert.

He was diagnosed with Parkinson's disease. His control over his body faltered. Toward the end he could neither stand nor walk unaided. He suffered several bouts of bronchitis, then renal failure, and finally came down with pneumonia. He lapsed into a coma for more than three weeks, then died.

Many of those who offered condolences pointed out that he had lived past the age of 89, and that was quite an accomplishment. Their attitude seemed to be that a person *ought* to die after reaching such an age.

"What else can you expect?" they seemed to be saying. It was if they believed that we are destined to wear out over the years, like a pencil that is used down to its nub.

Entropy

Strictly speaking, no one dies of old age. We die of disease or physical breakdown. We *expect* disease to overtake us or our internal systems to break down when we reach our "threescore years and ten . . . [or] fourscore" years. We expect to wear out.

What happens at the cellular level, inside our bodies, bears out such expectations—to an extent. Some biologists refer to a term from physics, *entropy*, to explain how we gradually succumb to the forces that are constantly trying to destroy us.

Entropy is the term used to represent the state of disorder in a system. Things run down. A highly organized marching band becomes a gaggle of individual men and women after the band's performance. The smoothly functioning muscles of a 20-year-old dancer lose their elasticity over the years. Someday the sun will go dark. The universe is winding down.

On the cellular level in our bodies, life is a constant struggle between the forces of entropy—the "thousand natural shocks that flesh is heir to"—and the body's powers of repair. Infection, ionizing radiation, and toxic chemicals all damage or kill our cells. Oxidation from free radicals and glucose browning are constantly on the attack. The body's defense systems work to repair damaged cells and replace those that are killed.

Over time, though, the body's defenses slow down. The damaging effects are not repaired as quickly as they once were; eventually, they are not repaired at all. We see the effects on the macroscopic level: Skin becomes wrinkled and coarse, bones become thin and brittle, teeth are lost, eyesight dims, muscles stiffen, arteries clog with cholesterol plaques.

We become more susceptible to infection and disease because we no longer have the vigorous inner defenses working to protect us. Our systems begin to slow down, or break down altogether.

We die.

The Fountain of Youth

Yet, ever since humans understood the inevitability of death, they have searched for a way of prolonging life. Gilgamesh, legendary king of the ancient Sumerian city of Uruk, sought immortality and failed. Ponce de León, Spanish explorer and adventurer, searched for the Fountain of Youth in 1513 and discovered instead the land of Florida (to which, today, many people retire to spend their "golden years").

In 1889 (the year that the Eiffel Tower opened) Charles-Edouard Brown-Séquard, a professor at the Collège de France, announced in Paris that injections of a liquid extract from the testicles of guinea pigs and dogs could rejuvenate a man physically and mentally. This seemed to confirm a long-held suspicion that male sex glands held the key to extending male vigor—and perhaps extending life span as well.

Others had proposed various nostrums for rejuvenation before Brown-Séquard, and still more would come up with similar elixirs afterward, but Brown-Séquard was a respected man of science, a professor no less. His announcement caused a sensation and sent entrepreneurs on both sides of the Atlantic scurrying to produce similar rejuvenating products.

Alas, it was all in vain. Brown-Séquard had fallen victim to the insidious placebo effect: He found the results he wanted to find. He was 72, and he wanted to feel younger. He tried his extract on himself and felt better. *Voilà!* Liquified testicles were the key to long life and (perhaps more important) a rejuvenated sex life.

It seems clear that much of the urge to find magical elixirs of rejuvenation centered on the male sex drive. Thus the interest in testicles as the secret to the Fountain of (more or less) Youth.

Brown-Séquard's "success" was taken so seriously by the public that clerics preached sermons condemning this unnatural and blasphemous attempt to circumvent the natural order of things. Accusations of "playing god" were hurled from pulpits. Even Sir Arthur Conan Doyle got into the game with a Sherlock Holmes story, "The Creeping Man," in which an aged professor uses extracts from monkey glands to make himself energetic enough to pursue a desirable young woman. Unfortunately, in Doyle's tale the extract works too well: The professor becomes more a brute ape than a man and is killed by a large dog.

In the early twentieth century, Viennese physiologist

Eugen Steinach discovered that the hormone testosterone is produced by the testicles. By grafting testicles or ovaries into castrated lab rats, he proved that the sexual behavior of the animal depended on the sex glands it had, not on whether it had originally been male or female. He even proposed that homosexual behavior is caused by a hormonal imbalance, a concept that is still being debated today.

This led Steinach to graft testicles taken from young rats onto old rats. He reported that the old-timers were rejuvenated and lived 25 percent longer than ungrafted males.

By 1916, Dr. Frank Lydston of Chicago went the next step. He grafted a slice of testicle from another man onto his own. Lydston was 54 at the time and reported a rejuvenating effect, "especially [in] physiosexual efficiency." Within months, he was grafting testicles onto patients whose own glands were abnormal or had been damaged.

While Steinach went on to pioneer studies of hormones and their effects, the idea of gland-grafting became popular and scandalous. The notorious "Doc" John Romulus Brinkley established a private clinic in Kansas in 1917 to graft goat testicles onto men who sought rejuvenation. Critics thundered against all grafting procedures, whether done by legitimate researchers or quacks. One of their strongest arguments was that the "grafters" were really offering to extend or increase their patients' sexual prowess—an obvious evil, in the moralists' view.

In the 1930s monkey glands became the "in" choice. Somewhat later Paul Niehans established an exclusive clinic in Switzerland, where he treated famous men such as Somerset Maugham, Noel Coward, and even the ailing Pope Pius XI with injections of extracts from sheep fetuses.

People found what they wanted to find, by and large. Eventually, though, even the most enthusiastic recipient of these "therapies" died, and the fad died with them. However, this quest for a Fountain of Youth laid the foundation for research into hormones, particularly the relationship of sex hormones to aging.

Why Do We Age?

Despite centuries of seeking a Fountain of Youth, aging and eventual death still seem inevitable. "Even in the midst of life we are in the midst of death," says the Book of Common Prayer.

Yet for the early part of our lives, we are growing physically and mentally. For the long middle years of our lives, we seem to hold our own, pretty much. Then it finally becomes noticeable: We are deteriorating, perhaps slowly, but inevitably.

Why do we age? Is it truly inevitable?

Biomedical researchers are divided in their opinions about the causes of aging. The majority lay the blame on evolutionary forces: We age because of entropy and an accumulation of genetic defects. A small but vocal minority claims that the cause is our cells' Hayflick limits: We age because our cells lose their ability to reproduce themselves.

Yet in the 1950s Hardin Jones, a biophysicist at the University of California–Berkeley, made the observation that a person's biological age (as opposed to calendar age) is determined to a large extent by the diseases the person experiences early in life. Reducing the incidence of childhood diseases not only extends the average life expectancy, it produces a population that is younger, biologically, than previous generations. Jones estimated that the generation born in the 1940s was physiologically five to ten years younger, at any given calendar age, than the generation born a century earlier.

We see this all around us today. Men and women at retirement age are still healthy and vigorous, not the shuffling old folks of earlier generations. Whole industries have grown over the past few decades to cater to the entertainment and education of "golden agers."

Yet even the most vigorous tennis-playing, bicycle-riding, adult-school-course-taking man or woman eventually weakens and dies.

Why?

121

The Evolutionary Explanation

The British immunologist Sir Peter Medawar, who won the Nobel prize in 1960 for his discovery of why the body rejects foreign tissue such as skin grafts (which led the way to successful organ transplants), expounded an evolutionary theory to explain why we age.

His theory started with an observation about the genetic defect that causes Huntington's disease, a neurological breakdown that usually does not show itself until early middle age. While its progress is slow, it is invariably fatal. The victim progressively loses control of his or her body and mind. Dizziness, spastic movements and facial tics, slurred speech, loss of memory, and bursts of rage are its symptoms. Worst of all, for most of the course of the disease the victim knows what is happening, and knows that nothing can be done to stop the deterioration.

Woody Guthrie, the famous folksinger and father of singer Arlo Guthrie, died of Huntington's in 1967 at the age of 55. He had been hospitalized for thirteen years.

The British geneticist J.B.S. Haldane pointed out that most genetic diseases tend to wipe themselves out, because they kill their victims before the victims have the chance to produce many offspring. People bearing a genetic fault that leads to a fatal disease will die early in life—unless the disease, like Huntington's, does not manifest itself until midlife or later, after the victim has had the time to produce children.

Medawar took Haldane's observation as the central point upon which he built his evolutionary theory of aging.

Darwin's theory of evolution hinges on the idea of natural selection.[16] Simply put, natural selection is the concept that organisms that live long enough to have offspring that survive and have offspring of their own are the species that survive from one generation to the next. If an organism fails to reproduce, its line dies away.

For example, if cancer killed off most human beings before they gave birth to children, the human race would have become extinct long ago.

Now, what about Huntington's disease and the evolutionary theory of aging?

Medawar observed that the longer an organism lives, the more damage it incurs from its environment. Entropy is constantly battering us. We are continually suffering damage to our DNA, mutations acquired from the harmful effects of free radicals, toxic chemicals, infection, ionizing radiation, and so on. Although our cells have repair mechanisms and can even replace some of the cells that have died, the longer we live, the more damage accumulates in our genes.

It is this accumulated damage that has been building up in the human genome for many thousands of generations that is the cause of aging in Medawar's view. Old age is the result of genetic garbage that has built up in our chromosomes. Entropy kills us.

---

[16]The word *theory* as used here is a term that scientists employ to denote a set of ideas, a blueprint or model that explains many observations and points the way to new understandings. It does not mean an unproven hypothesis or guess.

Mutations that do their damage early in life usually kill their host before he or she can have children. Those mutations eventually remove themselves from the gene pool. Mutations that do not begin to act upon us until later in life—after our child-rearing years—remain in the genome for millennia. The damage they cause, ranging from wrinkling our skin to causing Huntington's and Alzheimer's and cancer, are what we know as aging.

Our muscles stiffen and our eyesight fails not because our cells are wearing out, but because this accumulation of late-acting mutations interferes with the cells' natural ability to repair themselves. Genes that were once mutations have been passed down for thousands of generations so that they now are as much a part of our genome as the genes that code for the color of our eyes or the whorls of our fingerprints. Yet they are harmful genes, destructive genes, that lurk in our chromosomes through our early years and become active only later in life.

Find those genes, identify how they express themselves to cause the effects of aging, and we should be able to suppress their actions or replace them altogether. Aging, then, will evaporate like a bad dream.

11

Reversing Aging

*You may be offered an opportunity to become twenty
again—and remain so—for a far longer time than you
have yet lived.*

—MICHAEL FOSSEL

∞ THERE IS ANOTHER VIEW OF AGING, OPPOSED TO THE
entropic, or "accumulated garbage" concept of the evolu-
tionists. In this view, aging is caused by a specific process
in our cells, a mechanism that has been identified and
may be reversible.

"Aging is not a process that occurs passively as a result of living a certain number of years," maintains neurobiologist Michael Fossel of Michigan State University,
"but is a process that occurs as telomeric shortening
allows entropy to take you apart."

What is happening, in this view, is that as the various
sets of cells in the body approach their various Hayflick
limits, they cannot divide as quickly as they once did.
Eventually, they cannot divide at all. When cells are damaged or destroyed, it becomes difficult or impossible for
the body to repair or replace them.

Yet, as we know, some cells evade the Hayflick limit
and continue to reproduce indefinitely. Our germ-line
cells are essentially immortal; there is no Hayflick limit

for them. Cancer cells are also able to reproduce continuously.

Fossel and other researchers maintain that the telomeres, those caps at the end of each chromosome, are essentially the clock that governs cell reproduction. Each time a cell divides and reproduces itself, its telomeres shorten. Eventually, the telomeres become so short that the cell can no longer reproduce. That is what sets the Hayflick limit. Germ-line cells and cancer cells have the ability to rebuild their telomeres after reproducing and therefore can go on reproducing forever.

125

Might there be a way to coax the other cells of our bodies to get around the Hayflick limit and continue repairing themselves or reproducing just as they did when they were young? If there is a clock buried in the DNA that sets the number of times a cell can reproduce, can we find a way to reset that clock and keep the cell—and ourselves—eternally young?

Progeria: Death in the Fast Lane

Exactly the reverse is true for those poor souls afflicted with Hutchinson-Guilford syndrome. This horrifying genetic defect causes children to age so rapidly that they usually die before they are 20.

At birth, Hutchinson-Guilford babies appear normal. But very soon the effects of the syndrome—which is caused by a single mutated gene—become evident. The baby grows very slowly. It is always underweight, scrawny.

By the age of 10, the child is already turning gray and bald. The skin becomes wrinkled and spotted. In their teens, Hutchinson-Guilford victims begin to develop the ailments of 70-year-olds: Internal organs begin to fail, although the brain and mind remain normal. Many of the victims die of heart disease, doddering old men and women just out of their teens.

Hutchinson-Guilford syndrome is a form of *progeria*, the disease of accelerated, premature aging. There is another form of progeria that is perhaps even more horrifying, called Werner's syndrome. Werner's does not begin to manifest itself until puberty, when the hair begins to gray and the sex organs fail to develop properly. Werner's syndrome victims suffer from atherosclerosis and many forms of cancer; they generally live to be 50 or 60, when they usually succumb to heart disease.

Both forms of progeria are caused by the mutation of a single gene. It is a dominant gene; only one copy of it needs to be mutated for the disease to appear. Fortunately, progeria is very rare: Hutchinson-Guilford occurs once in about 8 million births; Werner's syndrome is slightly more common.

Progeria does not mimic all the effects of aging, however. Progeric boys seldom develop prostate problems; Hutchinson-Guilford sufferers of both sexes are not particularly vulnerable to cancer, cataracts, high blood pressure, stroke, or diabetes. Their brains do not deteriorate; they do not come down with Alzheimer's.

George M. Martin, a geneticist and pathologist at the University of Washington, has pointed out that the disease that most mimics accelerated aging is not progeria but Down syndrome. Down victims tend to lose their hair early in life and are prone to leukemia, diseases of the blood vessels (vascular diseases), and—inevitably—Alzheimer's. Some Down sufferers begin to show signs of Alzheimer's while still in their twenties; virtually every Down case develops Alzheimer's by the time he or she is 40.

If telomere shortening is actually the key to aging, the cells of progeric children should display much shorter telomeres than normal children's cells. That is exactly what has been found. The cells of progeric children have very short telomeres. "Compared with normal children," Fossel states, "the telomeres of Hutchinson-Guilford children *at birth* [Italics added] are like those of a ninety-year-old."

Among the many physical problems afflicting Down syndrome victims is their susceptibility to infection. The telomeres of their white blood cells and other cells of their immune systems shorten more readily than normal immune cells. In essence, their immune systems age more rapidly than normal.

Telomeres and Aging

If, as Fossel and other researchers suspect, telomeres really are the clock of aging, how do they work?

As we saw in Chapter 6, *telomeres* (the word is from Greek roots, meaning "end part") are the end caps on the chromosomes. Each human cell has forty-six chromosomes arranged in twenty-three pairs, so there are ninety-two telomeres in each cell. The telomere is made up of several thousand DNA bases and proteins that are bound to the DNA.

Telomeres prevent the threadlike chromosomes from sticking to one another. Without telomeres, the chromosomes would not be able to separate properly during cell division.

The very end of the telomere is folded into a hairpinlike structure. The telomere contains no genes. It does not carry the blueprint for proteins; its function appears to be to allow the genes along the rest of the chromosome to code for proteins properly. There are many long stretches along the chromosomes that apparently contain no genes, "nonsense" regions of monotonously repetitive bases whose purpose puzzles molecular geneticists. Are they accidents of history or do these seemingly arid regions perform functions that we do not yet understand?

Similarly, while telomeres are built of proteins and DNA, their DNA bases are tediously repetitive. In all vertebrate animals, the sequence of bases in the telomeres is exactly the same: thymine, thymine, adenine, guanine, guanine, guanine. Each cell in every fish, amphibian, reptile, bird, and mammal on Earth carries precisely the same telomeres: TTAGGG. In each telomere, this sequence (written T_2AG_3 in the scientists' shorthand) is repeated more than a thousand times.

In 1972 James Watson pointed out that each time a chromosome duplicates itself, it becomes shorter. What shortens, it was soon discovered, are the telomeres. The rest of the chromosome replicates itself exactly. But when a chromosome duplicates itself, it does not duplicate its telomeres. They are built afresh after duplication. Each time the chromosomes of our somatic cells reproduce, they rebuild their telomeres a little shorter than the telomeres had been before cell division.

By 1986, biologists discovered that telomeres from germ-line cells were longer than telomeres from other body cells. It was soon discovered that germ-line cells produce abundant telomerase, an enzyme that builds telomeres. Cancer cells also produce plentiful telomerase.

The implication is clear: In cells that age, the telomeres shorten with each cellular reproduction. In cells that do not age, the telomeres are rebuilt by telomerase after each reproduction.

However, there is some evidence that telomeres do not shorten every time a cell divides. Moreover, in October 1997 a team of researchers from the United States, Spain, and Canada reported on an experiment in which lab mice were genetically altered so that their cells produced no telomerase at all. The mice survived for six generations, even though their telomeres shortened, as expected, with each cell division. When exposed to oncogenes, the mice formed tumors, despite the lack of telomerase.

The researchers concluded that the mice must have some other way of maintaining the ends of their chromosomes. Simpler organisms, such as yeast, do so without telomeres; perhaps telomeres are not essential for cell division after all, nor is telomerase necessary for runaway tumor growth.

Yet in January 1988 researchers announced that they had extended the life span of human cells "indefinitely" in a laboratory experiment in which telomerase was added to the cells.

The researchers, from Geron Corporation and the University of Texas Southwestern Medical Center, inserted the gene that produces telomerase into the cells. They reproduced well past their Hayflick limits, giving powerful evidence that telomeres have a decisive influence on cellular senescence and may indeed be "the clock of aging."

Writing in the prestigious journal *Science*, biologist Titia de Lange, of Rockefeller University's Laboratory for Cell Biology and Genetics, commented, "The doubt [about telomeric influence on aging] has now come to an end with a report . . . describing direct evidence for a causal relation between telomere shortening and cellular senescence."

While most biologists do not accept the idea that telomeres are *the* clock of aging, and look to other causes, such as entropy, the Geron/Texas experiment shows what Fossel and his fellow telomerists have predicted: the "cure" for aging may be much simpler than anyone has expected.

The Long and Short of Telomeres

At conception, the fertilized egg begins with telomeres that are about ten thousand base pairs long. By the time of birth, after about fifty sets of cell divisions, the telomeres have shortened to some five thousand base pairs, or slightly more than eight hundred T_2AG_3 repeats. There is also a *subtelomeric region,* a sort of buffer zone between the telomere and the rest of the chromosome, that is about five thousand base pairs long.

The telomere, even with the subtelomeric region, is quite short when compared to the rest of the chromosome. The average human chromosome is about 130 million base pairs in length, twenty-five thousand times longer than the telomere at birth. The average gene is some 120,000 base pairs long, roughly twenty-five times the length of the telomere at birth.

Telomeres and Gene Regulation

Genes code for proteins. That is, each gene is a blueprint for the production of a certain protein. As a cell ages, it makes more of some proteins and less of others. In the geneticists' term, gene *expression* changes.

Some of the proteins that the genes produce are used to regulate the behavior of the genes themselves. It's as if you write yourself a list of instructions for carrying out a complex task. One of the most important aspects of this function is regulation of cell division. When cell division slows or stops altogether, we get the effects of aging. When cell division runs rampant, we have cancer.

At the end of each chromosome, nestled between the active genes and the subtelomeric region, is a set of genes called the *peritelomeric genes* (genes near the telomere). In young cells, these peritelomeric genes are covered by the telomere, like a cap pulled down over your ears, and remain inactive. But as the telomere shortens with each cell division, the peritelomeric genes are exposed. Once exposed, they are activated.

The peritelomeric genes code for proteins that suppress the activity of other genes. The production of enzymes in the cell is suppressed. *Catalase*, for example, is an enzyme that helps to protect the cell against free radicals. Its suppression leads directly to increased cell damage from oxidation.

This is the mechanism that some researchers consider to be the key to aging, the molecular reason behind senescence and eventual cellular death. Telomeres shorten with each cell reproduction. Their shortening eventually exposes the peritelomeric genes. Once these genes are activated, they suppress the normal functioning of other genes, which leads to a loss of the cell's ability to defend itself or repair damage.

Cells that produce telomerase constantly rebuild their telomeres and escape this vicious circle of eventual death. They continue to reproduce without aging. They are effectively immortal; they do not age.

131

The Long Good-bye

Thus, in the telomeric view, we age gradually as our individual cells reach their Hayflick limits and stop reproducing.

Our bodies consist of *germ-line cells*, which are immune to aging, and somatic cells, which age. But the *somatic cells* age at different rates, based largely on how often they reproduce.

The hardest-working somatic cells are the *stem cells*, which constantly replace blood, intestinal, skin, and other cells that are lost each moment of the day. You lose many millions of such cells every day; they must be replaced.

On the other hand, nerve and muscle cells do not normally reproduce themselves. If you lose them, they are not replaced. (Although, as we will see in Chapter 14, modern research is changing this.)

Each type of cell has its own Hayflick limit, which may very well be imposed by the rate at which the cell's chromosomes lose their telomeres.

If it becomes possible to rebuild the telomeres in somatic cells, as they are rebuilt in germ-line cells, it may become possible to halt—and even reverse—the effects of aging.

Remember, most bioscientists do not share this view of the central role of telomeres in the aging process. But before we address the arguments against the telomeric explanation, let us examine what might be possible if the telomeric explanation turns out to be correct.

This is a very big *if*.

Eternal Youth?

The telomeric explanation is based on the proposition that we age because our cells age, and our cells age because their telomeres shorten.

If that is true, if telomeres really are the cellular clock that determines the rate at which we age, then it might become possible to reset the clock—or even stop it altogether.

Researchers have essentially stopped the telomeric clock in human cells in a laboratory culture dish. They have found that adding telomerase to the cells rebuilds their telomeres after cell division and allows the cells to reproduce continuously without a Hayflick limit. Telomerase rebuilds the telomeres, like a molecular railroad construction crew laying down new track. The cells no longer age. But they don't become any younger than they were to start with. While young cells remain young, old cells stay old.

Although we do not yet know how to reset the clock of aging and make old cells young again, it appears that by feeding the cell telomerase, we can stop the clock from ticking.

How to get the telomerase into the cell?

Although most of the body's cells do not produce telomerase, the gene for its production is present in every one of the 100 trillion cells of your body. If we could get the cells to express that gene and make telomerase when we need it, we could halt the inexorable advance of aging. While not actually a Fountain of Youth, it would be at least a reprieve from senescence and death. (Again, assuming that the telomeric explanation of aging is correct.)

Yet the fact that most of our cells do not produce telomerase naturally gives pause. Is there a reason for this churlish behavior? There must be. More on that shortly.

For now, let us consider other ways to get telomerase into our aging cells. It is easy enough to inject the enzyme into cells in a culture dish—under controlled laboratory conditions. But the living human body is another matter altogether. Telomerase is a rather fragile chemical, easily broken up. The enzymes in the body's blood would destroy injected telomerase before it got the chance to start working.

Biochemists are clever people, though. It might become possible to develop an artificial, man-made form of telomerase, tough enough to withstand the body's defense systems while still maintaining the essential characteristics of natural telomerase: the ability to rebuild telomeres. Such a synthetic telomerase (or telomerase *analog*, to use the biochemists' jargon) could be injected into the bloodstream without being broken down; it could get to the cells to rebuild telomeres. Similar analogs have been developed for antibiotics, steroids, and other therapeutic drugs.

Most likely, telomerase analogs would be specific for

certain types of cells, rather than a broad-band medication that would affect the entire body. For example, a specific telomerase analog for resetting the clocks of blood vessel cells might reverse the effects of atherosclerosis, thereby preventing heart disease.

Another way of getting active telomerase into the body would be to engineer genes that produce telomerase. If your own somatic cells won't do the job, insert new genes that will. The engineered genes would produce telomerase once they have ensconced themselves in the nuclei of the body's cells.

Remember, however, that as yet we do not know the full molecular structure of telomerase, so it is currently impossible to engineer a gene to express the enzyme. We are in the position of wanting to build a factory for manufacturing an airplane—but we don't have the blueprint of the airplane yet.

Late in 1997, however, scientists at Geron Corporation, of Menlo Park, California successfully cloned the gene for human telomerase, and used the gene to extend the life span of human cells "indefinitely." So even though the structure of telomerase is not fully known, the gene for producing it can now be cloned and used to extend life span.

The question now is, how to get the gene into the cells? Here the already established techniques of gene therapy come into play. The engineered gene can be put into a virus or a liposome carrier, as described in Chapter 7.

Today bioscientists usually use viral carriers of altered genes *ex vivo*, outside the body. That is, they take a sample of cells from the person and put them in a culture dish. Then they insert the virus bearing the altered gene into the cells. After determining that the virus has done its job, and the cells are now expressing the new gene, the cells are then returned to the person from whom they were taken.

In the future, it will become possible to insert the virus directly into the patient, *in vivo*. However, the *ex vivo* technique has some advantages. It allows the scientists to make certain that the gene is working and the viral carrier is not causing any unwanted side effects before placing the new gene into the patient. In military parlance, you can "shoot-look-shoot."

The advantage of gene therapy, whether *ex vivo* or *in vivo*, is that once the altered gene has been inserted into the body it could "infect" other cells in the body. Once the new gene splices itself into the nucleus of a cell, that cell will start to produce telomerase. Its telomeres will be rebuilt constantly. That cell will not age.

And neither will the person to whom that cell belongs, if all the other cells in his or her body are similarly "infected" by the engineered telomerase gene.

This desirable fate will happen only if—and it's a big *if*—the cells produce enough telomerase to regrow the telomeres enough to stop the aging clock. Much research needs to be done to see if this will be the result. Too little telomerase and the whole procedure has been in vain. Too much and the result could be a tumor that refuses to stop growing.

There is another possible way to regrow telomeres in the somatic cells. Each of those cells contains a gene for telomerase. Why not learn how to activate that gene?

Researchers are hard at work on that approach. They are seeking a chemical compound that turns on the telomerase gene. This involves a dogged screening of thousands of possible candidate chemicals, but the goal is to be able to activate the body's natural telomerase production.

One of the fallouts from such research has been the discovery of *telomerase inhibitors*, compounds that slow or stop telomerase production. These are being tested against cancer. Slow or stop the telomerase production of cancerous cells and you slow the tumor's growth or kill it altogether.

If a telomerase activator is found, it could be modern science's elixir of youth. Assuming that the telomeric explanation is correct, a compound that encourages somatic cells to rebuild their telomeres could lead to perpetually youthful cells—and perpetually youthful people.

Since somatic cells naturally produce telomerase inhibitors, the "elixir" would be self-limiting. Apply the telomerase activator compound and the cells rebuild their telomeres, thereby resetting their aging clocks. Then the activator can be withdrawn until the cells begin to show signs of aging again.

The "elixir" will not wind the clock backward. A 50-year-old will not become 25 again. But he or she will not become 60, either, physiologically speaking. And a 20-year-old could remain 20 indefinitely.

Risks and Dangers

The nagging question remains, though. Each cell has its gene for producing telomerase, but in most cells the gene is not expressed. Why not?

Assuming that nature is neutral and doesn't care how long you live once you have passed the age where you can produce children, why don't our cells make telomerase and let us live forever? Or, at least, much longer than our "threescore years and ten . . . [or] fourscore" years?

The answer appears to be cancer. Telomerase inhibitors prevent cells from proliferating uncontrollably. They are part of the body's natural defenses against cancer—an important part.

In ancient myths and old fairy tales, when someone

gets her wish, it always comes with a catch. Cinderella may go to the ball, but she has to leave promptly at midnight. Achilles is granted great glory but only by dying young. King Midas is granted the golden touch, but it turns his beloved daughter into a golden statue.

Our cells do not express their telomerase gene, because cells that constantly rebuild their telomeres could—and demonstrably do—become cancerous.

But the history of humankind is filled with examples of our outwitting nature. When the Ice Age sent mile-thick sheets of glaciers across most of Europe and North America, we did not freeze in the cold. We learned how to use fire. When the oceans seemed an impassable barrier, we learned how to build ships that could cross them and how to navigate by the stars. Diseases that ravaged us for millennia eventually succumbed to organized scientific research, just as scientists are today working at ways to defeat AIDS and ebola.

Assuming that telomerase can stop aging, will we be smart enough to learn how to use telomerase without triggering deadly cancers? Don't bet against it.

12

Where We Stand Today

Nobody makes a greater mistake than he who did nothing because he could do only a little.

—EDMUND BURKE

∞ WHILE A SMALL ARMY OF SCIENTISTS IN LABORATORIES around the world labor to map the entire human genome and other investigators study the use of telomerase to stop cells from aging, many other avenues are also being followed in the pursuit of extending the human life span.

As we have seen, finding a way to evade the ravages of the years has always been of enormous interest, particularly among aging men who wanted to retain or renew their sexual vigor. Women have sought youthfulness too, usually because it helped them to attract or keep a mate. Thus was born the cosmetics industry (including cosmetic surgery), which to this day promises to make women appear young and attractive.

Eat Less, Live Longer?

The one sure finding that has come out of all this research is so commonplace that it may seem obvious: Good nutrition and exercise are vital to maintaining health. And healthy people live longer.

There is considerable evidence that childhood diseases shorten life. People who have suffered serious illnesses in childhood tend to have shorter lives than those whose childhoods were relatively disease-free.

But when it comes to *extending* life, there is abundant evidence that less is more. At least, for laboratory rodents.

Lab mice and rats have been carefully bred for hundreds of generations so that they are as identical, genetically, as scientific knowledge can make them. This is so that experiments on them can be conducted without being clouded by variations from one animal to another. To as great a degree as possible, in any particular strain of lab mice and rats, the individual animals are interchangeable. Results obtained on one set of them hold good for the entire strain.

Restricting the amount of food eaten by lab rodents slows their aging rate. No one understands precisely why this is so, but the evidence seems clear.

Usually, lab rats and mice are fed as much as they like. The animals themselves determine how much they eat. The food baskets in their cages are kept well-stocked so that there is always plenty of food available. This is called feeding *ad libitum*, more or less on the spur of the moment, like a standup comic's *ad lib* remarks.

In experiments to measure the effect of food intake on aging, some lab rodents were allowed to continue eating *ad lib*, while identical animals were put on a restricted diet: Their food intake was reduced by amounts ranging from 35 to 70 percent.

The dieting rodents lived an average of 25 to 40 percent longer than their free-feeding cousins. The longest life spans belonged to the animals that got the least food. Moreover, the dieters developed fewer tumors and other diseases. They looked and appeared youthful longer, and were better able to navigate mazes—an indicator of memory—than the fatter animals. They were more resistant to injected carcinogens as well.

As long as the dieters received adequate vitamins and minerals, the life-extending results occurred no matter whether their caloric intake was reduced in proteins, carbohydrates, or fats. The only thing that seemed to matter was that the overall amount of calories was reduced. The less they got to eat, short of starvation, the longer they lived.

What, if anything, does this mean for humans? It is hard to say. It seems obvious that restricting the lab rodents' food intake reduced the damage being done to their cells by oxidizing free radicals and glucose browning. In both rats and mice that underwent food restriction as soon as they were weaned from their mothers, puberty was delayed beyond the time of their *ad lib* eating cousins.

Adult rats slowed their reproduction rates, while mice stopped reproducing altogether.[17] Hormone levels and the rate of cell division were reduced for both species. Metabolic rate, the rate at which the animals burned their food to produce energy, went down when the food restriction first began but recovered to normal within a few weeks to a few months. Somehow the dieting lab animals burned just as much oxygen as normal, yet produced fewer free radicals in their cells.

Stress hormones went up, but this is hardly surprising to anyone who has done any serious dieting. The interesting point is that the animals appeared younger and healthier despite elevated levels of stress hormones. A certain level of stress, apparently, is not all that harmful; it might even have some beneficial aspects.

Food restriction experiments are now being tried on

[17]Human ballerinas and models who diet to the point of near-starvation cease their monthly menstrual periods.

rhesus monkeys, which are much more like humans than mice or rats are. However, while rodents live only a few years, rhesus monkeys live several decades. It will be some time before any definitive results can be expected from these experiments.

Some researchers believe that the life-extending effects of restricting lab rodents' food intake is not so remarkable, after all. They point out that mice and rats fed *al lib* are the lab rodent equivalent of couch potatoes. They eat too much and exercise too little. A restricted diet brings them to where they should have been in the first place, in this view.

For humans, the benefits of severely reduced diets are not readily apparent. About one-third of Americans are overweight, meaning they are 20 percent or more above the weight they should be, according to medical statistics. Yet Americans live as long as any other national population, on average, and Americans live longer after the age of 80 than any other nation's population.

Certainly, diet has its effects on longevity. But no one can today produce a diet that will extend your life.

Antioxidants

While it is impossible to link the *amount* of food we eat with predictable increases in longevity, are there certain foods or food supplements that can help us extend our lives?

Antioxidants are one obvious possibility. If free radicals damage our cells and age us through oxidation, why not ingest substances that are known to prevent the formation of free radicals in our bodies or break them up wherever they are found?

Vitamins A, C, and E are well-known antioxidants and can easily be obtained from fruits and vegetables or in larger doses from over-the-counter pills. But while they are of some help in avoiding cancer and heart disease, there is no valid evidence that they slow the aging process, even in megadoses.

In animal tests, antioxidants have shown only slight improvements in longevity at best. Many tests have produced no noticeable effect.

Vitamins are forms of enzymes that our bodies need but do not produce for themselves. Instead, we obtain these needed chemicals from the foods we eat. Enzymes are catalysts; they help certain chemical reactions to proceed in the body. Most vitamins have more than one effect. Vitamin A, for example, is crucially involved in the development of bones and eye pigments necessary for good vision. Vitamin C is not only an antioxidant; its major function is to help the body to produce collagen.

What about other antioxidants, such as *catalase* or *superoxide dismutase*? You may feel better by taking them, but they won't do you any good. For one thing, the cells of your body already produce these enzymes plentifully. For another, when you take them orally, these chemicals are broken up by your digestive system long before they can be of any help as antioxidants. Yes, you may find them on sale at health food stores. But their only effect on you will be psychological.

Melatonin

In 1995 melatonin became something of a medical fad. It was touted as a sort of miracle drug that helped people to sleep better and lose weight, prevented heart disease, cancer, Parkinson's, and Alzheimer's, invigorated their sex lives—and slowed down the aging process.

Virtually all of the evidence cited in favor of melatonin has been anecdotal: that is, reports by individuals of their reactions to the hormone. Such reports are notoriously subject to the placebo effect: The patient finds what he or she wants to find. Hard, cold laboratory results have been slim.

The pineal gland, which is located at the base of the brain, atop the brain stem, produces melatonin naturally. The pineal secretes five to ten times as much melatonin at night as it does by day. While melatonin apparently helps the body keep track of the day/night cycle, it actually *inhibits* the production of the sex hormones, testosterone and estrogen.

The pineal's production of melatonin reaches its peak at about the age of 5, then slowly wanes. If melatonin is involved in human aging, this is puzzling, because the human body does not begin to age (as we understand aging) until just before the onset of puberty.

In laboratory tests, giving rodents extra melatonin provided some antioxidant effects and lowered cholesterol. It also inhibited ovulation, which brings up the possibility of a melatonin-based contraceptive pill someday.

While some researchers have claimed that the lab mice which received extra melatonin showed 20 percent gains in life span, others have countered that the particular breed of mice used in these tests have a genetic defect— they produce no melatonin at all naturally. Yet, even without getting melatonin from the researchers, these mice live normal life spans. So if melatonin is involved in extending life span, how come mice that produce no melatonin at all live to the same "ripe old age" as mice that do produce melatonin naturally?

In humans, melatonin has been used as a sleep inducer, particularly for travelers subjected to jet lag. To add my own bit of anecdotal "evidence," melatonin has had little or no such sleep-inducing effect on anyone I know who has tried it.

The best that can be said about melatonin at this stage

of knowledge is that its antiaging effects are unproved. Moreover, possible side effects from melatonin can be severe, so consult your physician before taking this over-the-counter drug.

DHEA

Dehydroepiandrosterone (DHEA) is a steroid that is secreted by the adrenal glands. In fact, it is one of the most abundant hormones in the human body. Yet its function is unknown. The body can convert it into testosterone or estrogen. Its production can be suppressed by stress. Otherwise, no one knows what it does or why it exists.

Unlike other steroids, DHEA's level in the blood decreases rapidly after puberty, which may be a clue that it is somehow involved in aging.

In laboratory tests, mice treated with DHEA (which they produce naturally in extremely small amounts) live longer than mice that do not receive DHEA. But why and how are unknown. There is some evidence that in lab rodents DHEA suppresses obesity, tumor formation, diabetes, heart disease, and some forms of immune diseases. These results may be due to the fact that DHEA suppresses appetite so that the treated mice are in effect put onto a restricted food regime.

In Europe, DHEA is sometimes prescribed to ease the pain of gout and the discomfort of menopause. In the United States, DHEA is considered a nutritional supplement, and is not a controlled drug. Although it is available over the counter, the purity and contents of the pills are not regulated by the Food and Drug Administration.

As the Romans said, *Caveat emptor* ("Let the buyer beware").

Deprenyl

Deprenyl is a drug that has been used to control the trembling limbs and other symptoms of Parkinson's disease.

In lab tests, deprenyl has been shown to increase the life spans of elderly rats by 10 to 40 percent. The rats' sexual activities regained youthful vigor, and their memories improved (as shown by their abilities to remember the shortest routes through mazes). Since the drug apparently did not reduce the rats' appetites, the antiaging effects do not seem to be associated with restricted food intake.

Deprenyl's "secret" may be that it is an efficient antioxidant. When rats were sacrificed (lab jargon for "killed") after three weeks on deprenyl, their brains showed high levels of antioxidants, particularly in the regions where Parkinson's would wreak most of its damage.

While these results are encouraging—even exciting—it is important to understand that they have been obtained only in aging male rats. There is a long way to go.

Human Growth Hormone

One protein that seems to have a measurable effect in extending life span is *human growth hormone* (hGH), also known as *somatotropin*. Human growth hormone is secreted by the pituitary, a tiny gland located at the base of the brain, not far from the pineal. The pituitary is known as the master of the endocrine system, because it regulates the secretion of hormones by the body's other ductless glands, such as the thyroid, adrenals, gonads, and so on, which, in turn, secrete the hormones that regulate the body's metabolism.

Human growth hormone has many vital functions,

ranging from building bones and muscles to strengthening the immune system and helping to heal wounds. Dwarfism is a consequence of lack of hGH, and abnormally small children are treated with it to help them to grow closer to normal size.

While growth hormone is secreted in large amounts by adolescents during sleep, in adulthood its production diminishes, often as much as 10 to 15 percent every ten years in men. Women maintain higher levels until they reach menopause, then their hGH production rapidly declines. Researchers have not missed this obvious hint that slowdown in hGH may be related to aging.

Low levels of hGH are associated with loss of muscle leanness and accumulation of fat. Worse, as the individual's weight increases, hGH levels drop more, setting up a negative feedback loop. Injection of hGH improves muscle strength and leanness, apparently helping the body to build protein instead of fat.

Incidentally, menopausal women on hormone replacement therapy get a side benefit: The estrogen they take stimulates natural production of growth hormone, even in elderly women. Testosterone does the same for men.

Dr. Daniel Rudman of the Medical College of Wisconsin tested hGH's antiaging possibilities with a group of male volunteers in their sixties and seventies. Half the volunteers received no hGH injections; they were the control group. The others received hGH injections three times a week for six months so that their hGH levels were returned to the amounts they had as young men.

While the men in the control group showed the normal deterioration of muscle, bone, and organs expected for men of their age, those who received the hGH injections not only stopped aging—in some ways, their aging was reversed.

They put on new muscle mass. Their skin increased in thickness by almost 10 percent. Internal organs, such as the spleen and liver, also gained mass. Some of the deteriorative effects of aging had been stopped and even turned around.

To make certain these effects were due to hGH, the researchers stopped the injections. The "youthful" group began to age normally once again.

It is far too early to be certain, but if there is a single elixir of youth, human growth hormone might be it. Still, hGH needs to be tested over long time spans, and its possible side effects must be tracked down. It is known, for example, that overly large doses of hGH can cause or aggravate hypertension, lead to diabetes, enlarge the heart, and affect the joints.

147

However, the preliminary results to date give reason for cautious optimism. Commenting on the future possibilities of various therapies for aging—from food restriction to hGH—molecular biologist John Medina, of the University of Washington School of Medicine puts it this way:

> *What this means is nothing less than a bombshell. There are active researchers in the field today who think that we may soon have protocols that could double or even treble normal human life spans.*

Cryonics

How soon is "soon"?

We can see that many lines of research are probing the unknowns of human aging. Very likely, in a few decades, biomedical science will be able to extend our life spans for centuries.

But what about men and women who cannot wait a few decades? What about people who are facing imminent death and will not live long enough to see the break-throughs that will lead to immortality?

There is a possibility that they might evade their fate. A slim possibility, to be sure, one fraught with unknowns. Yet a possibility does exist.

Cryonics.

Pharaohs of ancient Egypt had themselves embalmed and entombed in the belief that their bodies (and possessions) could be preserved for a life after death.

Today a small group of people believes that a person who is clinically dead can be frozen in liquid nitrogen and preserved indefinitely to await a day when he or she can be revived, brought back to health, and begin a new life.

This concept—called *cryonics*—was originally championed by Robert C. W. Ettinger in his 1964 book *The Prospect of Immortality*. A teacher at Wayne State University and Highland Park College, both in the Detroit area, Ettinger has degrees in physics and mathematics. In 1967 Ettinger and his followers formed the Immortalist Society, and in 1976 they founded the Cryonics Institute in Michigan to offer the facilities and services for freezing bodies and maintaining them in long-term storage in liquid nitrogen (see Liquid Nitrogen, below).

LIQUID NITROGEN

Nitrogen is gaseous under normal conditions. It comprises nearly 80 percent of Earth's air.

Cool nitrogen to −320.8° Fahrenheit (−196° Celsius) and it becomes a liquid. The technology of dealing with ultracold materials such as liquid nitrogen, liquid oxygen, and liquid hydrogen is called *cryogenics*. It is a development that was greatly enhanced by the Space Program's need for such liquified gases.

As of this writing, the Cryonics Institute has a membership of 180 and houses twenty frozen bodies.

Other organizations have arisen as well, such as the Alcor Life Extension Foundation in Arizona.

The idea behind cryonics is that when a person is declared clinically dead, the body can be frozen in liquid nitrogen and preserved indefinitely. The body will not decay while immersed in the cryogenically cold liquid nitrogen. In time, biomedical science may learn how to cure the ailment that caused death and revive the patient, restoring him or her to life.

There have been some grisly tales associated with some cryonics establishments: tales of freezer facilities that failed and allowed the bodies to rot, tales of people who chose to have merely their heads frozen on the assumption that some time in the future it will be possible to attach their heads to a new body.

Such tales aside, cryonics seems no more unreasonable than the ancient pharaohs' preparation for an afterlife. Basically, those who choose to have their bodies frozen are making a bet. They are betting that: (1) their cause of death can eventually be cured; (2) they can be revived after storage in liquid nitrogen; and (3) their frozen bodies will be faithfully preserved until (1) and (2) can be accomplished.

If they lose the bet, so what? They are already dead.

The major question is: Can a human body be revived after being frozen? No one has accomplished that yet.

A large part of our body mass is water, and when frozen, water ice is less dense than the liquid form. This is why icebergs float, rather than sink. When the ice in frozen cells thaws back to liquid water, the cells tend to swell and burst.

Fish and frogs can be frozen and thawed easily enough. No large mammals have been successfully thawed, although human embryos, sperm cells, skin,

149

bone, blood cells, and bone marrow have been successfully thawed after freezing. Laboratory rat hearts have been frozen in liquid nitrogen, then thawed and started beating again.

Cryonics is undoubtedly the most "heroic" of techniques to stave off the inevitability of death; it comes into play only after the patient is declared clinically dead. It is a bet on the future. To most people, it seems weird—even macabre—at first. But it is a possibility that should be examined on its merits.

Prosthetics

In the motion picture *Miracle on 34th Street*, when asked his age, Kris Kringle replies, "I'm as old as my tongue and older than my teeth."

The debilitating effects of aging can be countered, to some extent, by replacement parts for your body. This concept goes back to the seventeenth century and the invention of eyeglasses to assist fading eyesight. Or perhaps we should include canes, crutches, and the crude prosthetics such as wooden "peg legs" or the hooks that replaced lost hands. Certainly, every schoolchild remembers that Paul Revere made a set of wooden false teeth for George Washington.

Over the centuries, we have developed hearing aids and much improved prosthetic limbs. In 1967 the first successful heart transplant operation was performed.

Today it is possible to replace knees, blood vessels, corneas, bones, and ears. Cosmetic surgery rebuilds damaged faces and improves noses, chins, cheeks, breasts, buttocks, and tummies. Metal and plastic joints are available for hips, elbows, shoulders, and fingers, as well as knees. Bladder stimulators help paraplegics and incontinents. Dental implants replace lost or damaged teeth. Metal rods replace shattered bones, and prosthetic bracing strengthens and straightens damaged spines. Pacemakers control heart rhythms. A *brain pacemaker* was approved for clinical use by the FDA in 1997; it helps to control the shaking limbs of Parkinson's and other conditions that cause tremors.

Prosthetic replacement parts can help people to overcome some of the damage that comes with age or accident. But they will not halt or even slow aging.

That capability lies elsewhere.

Genes, Telomeres, and Growth Hormone

If time doesn't get you, chance will.
—STEVEN N. AUSTAD

∞ TIME FOR A REALITY CHECK.

We have two schools of thought concerning human aging and death.

The *entropic* school says that we accumulate genetic damage over the years, until finally our genes have amassed so many defects that they can no longer repair themselves adequately. Some 7,000 to 100,000 genes are involved in the many different aspects of aging, according to geneticists. As these genes fail, the cells function poorly or die. Our bodies age and stiffen. We are less able to fight off infections. Our systems—heart, kidneys, liver, and so on—slow down or fail altogether.

Eventually, we are like the doddering old man described by Shakespeare in *As You Like It:* "Sans teeth, sans eyes, sans taste, sans everything."

The inevitable result is death—unless biomedical scientists succeed in developing therapies such as human growth hormone injections that can stop or even reverse these effects of aging.

The *telomeric* explanation for aging, on the other hand, portrays the telomeres that cap each chromosome as the clock of aging. Each time a cell divides, the telomeres at the end of each chromosome shorten. When the telomeres get so short that the peritelomeric genes are exposed, they begin to inhibit the cell's genes and lead to all the effects of aging that the entropic argument describes above.

Germ-line cells and cancer cells avoid aging because they produce telomerase, which rebuilds their telomeres after each cell division.

Which explanation is correct? The answer is literally a matter of life and death.

Hayflick Limits

The telomeric argument is based on the fundamental observation that each type of cell has a limit on the number of times it can divide: its *Hayflick limit*. But the entropic argument counters that aging is actually something different from the cells' inability to reproduce.

Take muscle cells, for example. After about age 30, we lose roughly 1 percent of muscle strength per year. This is not because the muscle cells hit their Hayflick limits; muscle cells do not divide at all after being formed in development. They may grow larger with certain types of exercise, but no new muscle cells are produced, no matter how many hours you spend in the gym. The loss of muscle strength is caused by the individual cells' weakening over the years and the death of some of those cells.

Besides, say the entropists, cell division in a laboratory culture dish is not necessarily the same thing as cell behavior *in vivo*. In the living body, cells are bombarded with chemical messages from other cells; life is a constant give-and-take among the 100 trillion cells of the human body, not the controlled artificial environment of a laboratory culture dish. The Hayflick limit, the entropists believe, is linked to the body's defenses against cancer and has little or nothing to do with aging.

The idea here is that evolution has selected a cruel choice for us. When a baby is conceived, the fertilized egg must grow from a single cell into a full human body of some 100 trillion cells. Fetal cells have the longest Hayflick limit; they *must* be able to reproduce if the fetus is to become a normal, healthy baby. But once the body is formed, further rapid cell reproduction could lead to cancer, hence cell division is inhibited.

If cell reproduction continued at the prolific pace it accomplishes in the womb, we would all die of cancer very quickly. Natural selection would have wiped out anyone whose cells behaved that way before they could have many (or perhaps any) offspring. What remains are individuals whose cells reproduce more slowly, once the body is fully formed.

The downside of this choice is that, in avoiding rampant cancer, we face aging and death. But to the evolutionary force of natural selection, what is important is living long enough to bear offspring. That is the key to natural selection: Creatures that successfully produce offspring survive; those that do not become extinct. It is that simple, that stark.

What happens to the individual after it has produced offspring is not important, as far as natural selection is concerned.

Perhaps there is an exception to that last statement. Human babies need long years of nurturing and protection before they are able to mate and reproduce. For us, the ability to live long enough to see our children into puberty is vital. Despite the occasional myths of human babies reared by wolves or apes, babies do not survive the death of their parents unless some other humans take care of them.

Telomeres and Progeria

To the entropists, telomeres are interesting because they may show how to limit or prevent cancer. By slowing and eventually stopping cell reproduction, telomere shortening may be the body's first line of defense against cancer. Where the telomeric researchers want to learn how to produce telomerase to renew cells' ability to reproduce, the cancer researchers want to learn how to restrict telomerase production in cancer cells and thereby kill them.

155

To the argument that cell division is not the same thing as aging, the telomerists point to progeria. Progeric patients wither and age with heartbreaking rapidity, and they are born with extremely short telomeres.

The entropists counter that progeria is not the same as normal aging, it merely mimics some of the effects of aging. Biologist Roger Gosden points out that about 10 percent of "serious inborn genetic diseases" show at least some symptoms of premature aging. "Aging is not just an unpleasant supper served up late in the day," according to Gosden, "but is never very remote and should concern us all."

Whether they are truly the key to aging or "merely" the key to cancer, telomeres will be under intense investigation by bioscientists. If the telomerists are correct, the elixir of eternal (or, at least, greatly extended) youth is telomerase.

Worst Case Scenario

Recognize, however, that even if telomerase is the stuff of the Fountain of Youth, it cannot reverse the effects of aging. All the telomerase on Earth will not grow a third set of teeth for you.

What telomerase can do (if it works) is to slow or even

stop aging. If you are 30, you can remain physiologically that age as long as you can get telomerase to reset the aging clock(s) in your cells.

If it works.

Even if the telomerists are on the wrong track, as far as aging is concerned, the knowledge to come from the Human Genome Project will most likely lead to new ideas and new techniques on how to strengthen the body's cellular and genetic defenses against the ravages of aging.

There are genes in your cells that *can* grow you a third set of teeth, if only we knew which genes they are and how to activate them. That is what the Human Genome Project is offering us.

Already, researchers have discovered the genes involved in diabetes, cystic fibrosis, and other genetic disorders. As I am writing this, biologists at the National Human Genome Research Institute have reported identifying the gene responsible for Parkinson's disease. It also appears to be implicated in Alzheimer's.

And discovery of genes that influence aging in yeast and nematodes are already being linked to human genes that code for insulin regulation and influence glucose browning.

Once molecular biologists have mapped the entire human genome and begin to understand how acquired mutations diminish or destroy the cells' ability to repair themselves, then they will be on the road to understanding how to renew those abilities; how to return cellular function to a youthful, vigorous condition.

In the meantime, other scientists are pursuing the therapeutic uses of human growth hormone and investigating the possibilities of other treatments, such as DHEA and antioxidants.

Human growth hormone seems to play a pivotal role in the aging process. If the preliminary experiments done to date are borne out, hGH replacement therapy may become the key to stopping and even reversing the effects of aging. Will this extend the human life span? Unknown, but at the very least, we may have the opportunity here to maintain a relatively youthful, vigorous life right through to the final years.

It seems more likely that instead of a single elixir or Fountain of Youth, scientists will present us with an array of treatments and medications, ranging from gene therapy and telomerase to antioxidants and hGH, that will allow us to live longer, healthier, more active, and youthful lives. Centenarians may look, feel, and behave like 30-year-olds. The world record for longevity may be extended gradually from 120 to 200 and beyond.

Eventually (and that means anywhere from ten to a hundred years from now), real breakthroughs may be made and human life spans extended for many centuries or even more.

In the meantime, there is another possibility that we should examine: regeneration.

14

Regeneration

I believe that, not too far into the next century, we will be able to regenerate a number of vital tissues.
—DAVID L. STOCUM

∞ AT THE 1997 MEETING OF THE AMERICAN HEART ASSOCI-ation, two research teams revealed that they had used gene therapy to coax the body to grow fresh arteries as natural bypasses for clogged blood vessels.

Scientists from St. Elizabeth's Medical Center in Boston reported that they had injected twenty patients with a laboratory-produced version of the human gene that makes a protein, *vascular endothelial growth factor*, which spurs the growth of blood vessels in human fetuses. Sixteen of the twenty patients, who suffered from severely clogged arteries in their legs, grew new arteries that carry blood around the clogged vessels. The constant pain that had afflicted them has been relieved, and they have been spared the need for leg amputations. Several of the patients have even started to grow toenails again.

The second team, from Brigham and Women's Hospital in Boston, is using similar gene therapy to make coronary bypass surgery work better.

Early in 1998 researchers at the Fulda Medical Center

in Germany reported that they were able to trigger the growth of new coronary blood vessels by using a genetically engineered human growth factor, FGF-1. The growth factor was injected into twenty volunteers as they underwent conventional heart surgery. Within four days, all twenty patients began to grow new blood vessels.

By learning how to get the body to generate new blood vessels on demand, such work will eventually eliminate the need for bypass surgery. Why cut blood vessels from various parts of the body and staple them into the heart's arterial system when you can get the body to grow new arteries on demand?

Four months earlier, a pair of Harvard researchers announced that they had grown replacement organs—including hearts, kidneys, and bladders—for lab rats, rabbits, and sheep. They used the animals' own cells as the starting material, grew the new organs in the laboratory, and then implanted them surgically into the animals.

The two scientists who made this announcement—Anthony Atala and Dario Fauza—pointed out that one of the earliest uses of their work could be to correct birth defects while the baby is still in its mother's womb.

They gave the example of a baby with a malformed trachea. Surgeons could take some of the fetus's cells, grow a new windpipe for the baby, and have it ready as soon as the baby is born. Since the replacement organ is made from the baby's own cells, there would be no problem with tissue rejection, as there is with today's organ transplant procedures.

Their example struck me in the heart. One of my grandsons was born with a misconnected windpipe and nearly died before he was 12 hours old. Only masterful surgery saved his life. Now he's 5 years old, as vigorous and vital as only a little boy can be. Luckily, he did not need a transplant, only some corrective "plumbing" work to put his windpipe in order.

Soon, though, newborns who do need a transplant will

get one—built from their own cells, so there will be no danger of rejection.

When an organ is transplanted from another person, the recipient's immune system recognizes the transplanted organ as foreign tissue and attacks the organ, just as it would attack invading microbes. Successful transplants depend on drugs that suppress the recipient's immune system, a ploy that has considerable danger attached to it.

A friend of mine lost function in both his kidneys, due to a disease inherited through his mother. His father donated one of his kidneys to save his son. The transplant worked, but the man has spent the past ten years on immune-suppressive medications. He's alive, but his lifestyle is greatly slowed down, and he is constant danger from the slightest infection or cold virus.

If the transplant tissue were made of the recipient's own cells, the problem of rejection by the immune system would no longer exist.

Within five years, we should see the field of organ transplants begin to change into the field of organ regeneration. There will be no need to hunt for organ donors, no need to wait for some unfortunate accident victim's death so that you can get yourself a healthy heart.

And it won't stop there. If new organs can be grown in the laboratory, why can't you grow new organs or limbs in your own body? Why have coronary bypass surgery if you can grow new coronary arteries right there in your chest? Why get a prosthetic knee if you can build a new one out of your own cells? Why undergo cosmetic surgery if you can grow smooth young skin and tighter muscles for yourself?

E Unum, Pluribus

Once you were nothing more than one, single little cell, a fertilized egg.

Your one cell divided into two, and they divided, and divided and divided again—about fifty cycles of cell division took place. Your cells changed as they divided. Some became muscle cells, some became nerve cells, some became skin, heart, eyes, hair, fingernails, arms, legs, and all the different parts of your body.

Now you are a smoothly functioning individual of some 100 trillion cells, all of them the offspring of that one microscopic cell that was the original you.

Your cells contain all the information, in their genes, to build another copy of you. That is the basis for cloning. While moralists heatedly debate the prospects of cloning human beings, there is something much more important and valuable involved in the fact that almost every one of your cells contains the complete genetic blueprint of all of you.

Almost every cell of your body holds, in its nuclear DNA, the blueprints for every part of you: your skin and bones, your brain and nervous system, every organ in your body. If only we could learn how to *use* that information properly, we could grow new organs, limbs, skin, muscles, and any other part of the body when we need to.

Why buy a new automobile when you love the one you have and all it needs is a new fuel pump and a fresh set of tires? Why clone another whole copy of you when what you really need is a new heart and a fresh pair of kidneys?

Remember, a clone of yourself will be another person— and a baby, at that. You can grow a fully functioning heart in nine months. Perhaps it will only be baby-sized, but there may be ways to accelerate its growth to adult size quickly.

The Perfect Artificial Heart

As I mentioned in Chapter 4, in the early 1960s I was the manager of marketing for the Avco Everett Research

Laboratory (AERL), a major industrial research establishment in Massachusetts, a few miles outside Boston.

Although AERL was devoted to studying the physics of high-temperature gases (which led to our being called "hot air specialists"), the founding director of the lab, physicist Arthur Kantrowitz, often worked with his brother Adrian, who was a leading cardiovascular surgeon in New York. The brothers developed an early heart pacemaker and a bladder stimulator that helped save paraplegics from life-threatening bladder and kidney infections.

They also worked on a form of artificial heart, a plastic pump that would not replace the natural heart but would take over a large part of a damaged heart's workload and pump blood through the vascular system.

During Lyndon Johnson's administration, the National Heart Institute (as it was then called) inaugurated a sizable program to develop a complete artificial heart that could totally replace a failing natural heart. AERL won one of the NIH's contracts, although our lab continued to work on partial—and even temporary—heart assist devices, rather than a total replacement device.

It was during one of the brainstorming sessions, early in the program, that the question arose: What would be the *perfect* replacement heart? Unrestricted by limitations of technology or funding, what would be the best replacement for the natural heart that we could imagine?

I thought at the time that the perfect replacement would be a new natural heart, grown within the body itself, out of its own tissues. There would be no problem of tissue rejection, as there were with heart transplants. There would be no need for surgery. There would be no problem of figuring out how to make a mechanical pump respond correctly to the delicate chemical and electrical signals that regulate the natural heart's moment-to-moment performance. The replacement would actually be another natural heart, grown just the way the original one was.

That was far beyond anyone's ability to do more than dream of in the 1960s. It was the stuff of fiction, and many years later I wrote a novel on the subject.[18]

By the 1990s, however, biomedical research was catching up to the brainstorming dreams of the 1960s. Research on tissue regeneration is very real, and—as the Harvard researchers' announcement shows—it is beginning to show results.

The Limits of Specialization

The human body has some regenerative powers. We are constantly replacing skin cells and the red and white cells in our blood. The liver has considerable regenerative power, although our other organs do not. Muscle cells and nerve cells normally do not regenerate.

Some animals are much better at regeneration than we are. Many reptiles can replace a lost tail. Newts and salamanders, amphibians both, can regenerate not only lost tails and limbs but heart muscle and retinal nerves too.

Although each of us grew from a single cell, we lose

BABY TALK

There are several different technical terms for a baby developing in its mother's womb, depending on the stages of development. A fertilized egg cell is called a *zygote*. As the cell divides and grows, it is referred to as an *embryo*. Once the human body form is apparent, after the third month of pregnancy, the embryo is called a *fetus*.

---

[18]*Brothers*, published in 1996.

the ability to grow new limbs and organs once our original equipment is in place.

Bioscientists say that the early zygotic cells are *totipotent*. That is, each of those cells has the capacity to differentiate into all the various types of cells of the adult, from toenails to the "little gray cells" of the brain's cerebral cortex. As the zygote grows into an embryo, its cells become more and more restricted. They are *multipotent*, which means they have the capacity to produce a large number of different kinds of cells but not all the different kinds that the earlier totipotent cells are capable of producing.[19]

Once the embryo's cells differentiate into specific types of cells—heart, liver, nerve, bone, and so on—they lose their pluripotency. Those cells have specialized, and they will remain specialized. A muscle cell, for example, will not change into a nerve cell. The cells of the lungs will not suddenly transform themselves into red blood cells.

You can see why. Imagine trying to run a business organization where each day each employee decided which job he or she wanted to do. Some flexibility is possible in business, but such total lack of organization would lead to nothing but chaos.

In the body's development, once the cells have specialized (or *differentiated,* in the technical jargon) they cannot change into another type of cell.

However, each and every cell still carries in its nuclear DNA the genetic blueprint for the entire body. All the genes reside in each cell, although most of them are restricted from acting once the cell has differentiated.

Once the cell has differentiated, most of its genes are turned off. Only the genes that are necessary for the cell's specialized job remain active. The rest remain dormant forever.

Unless bioscientists learn how to activate them.

---

[19]Multipotent cells are also called *pluripotent.*

Act Like a Child

There is a poetic aptness here. To regenerate a damaged or diseased organ, we must learn how to make its cells behave as if they were multipotent. To regain youthful vigor, we must regress at least some cells to their embryonic state and make them behave as if they were back in the womb.

Scientists are pursuing three avenues of investigation leading toward this goal: studying the body's stem cells, which produce new blood, skin, and other types of cells all the time; studying organs of the body that regenerate naturally, such as the liver; and studying animals, such as the amphibians, that are masters at regeneration.

We have seen that stem cells are generating new cells all the time: Skin and blood cells, in particular, are constantly being produced by the body's stem cells. Stem cells behave like multipotent embryonic cells to a degree.

The somatic cells were long thought to be unable to generate new copies of themselves; neurons (nerve cells) in particular were considered to be once-only elements, in place for life and irreplaceable if damaged or destroyed.

Recent research has suggested that this is not so. "Multipotential stem cells have now been identified in some nonregenerating adult tissues, such as the brain," says David L. Stocum, biology professor at Indiana University–Purdue University.

If stem cells can be found and stimulated to produce fresh somatic cells on demand, regeneration of faulty organs or lost limbs may become possible. Even lost brain cells may be replaced.

Since skin cells are constantly replaced, it may become possible to control the rate at which their stem cells work, to stimulate faster and better wound repair and perhaps erase the wrinkling and coarseness that comes with age.

This form of regeneration may open a new era in cosmetic surgery.

However, most somatic cells are not replaced easily (or at all) in the body. Can we learn how to make them regenerate themselves?

To do this, we must uncover ways to *de*differentiate such cells; that is, make them behave as if they are multipotential embryonic cells again, capable not only of copying themselves but of producing other types of cells as well.

"Liver, Come Back to Me"

The liver is one organ that does regenerate itself after injury. This was apparently known to the ancient Greeks, who regarded the liver as the body's most important organ, the key to human emotions (much the way we today regard the heart, at least in poetry).

It seems that the ancient Greeks even understood that the liver can regenerate itself. In the myth of Prometheus, that demigod is chained to a rock and an eagle eats out his liver every day, while the liver regrows overnight, so the torture can be repeated the next day. (Prometheus was sentenced to this terrible punishment for the heinous crime of giving fire to the human race. The Olympian gods did not like the idea of man gaining such power.)

In modern laboratory experiments, large sections of the livers of rats and mice have been removed surgically. The liver grows back to its normal size in a week or less. Its cells multiply until the organ has reached its proper size and then they stop proliferating. Studies in dogs, monkeys, and even humans have also shown this remarkable regenerative quality. On the other hand, when the liver from a large dog is transplanted into a much smaller dog, the liver gradually diminishes until it is the same size as the smaller dog's original organ.

It is clear that the liver cells are receiving chemical signals from the cells around them that regulate the way they multiply—or die apoptotically, where the liver shrinks. Investigators have determined that liver regeneration does not depend on the presence of stem cells, such as the bone marrow stem cells that produce blood cells. And—most remarkable of all—the liver cells go right on performing all the necessary functions of the liver, even while they are multiplying furiously.[20]

A single *hepatocyte* (a liver cell) can go through thirty-four cell divisions, thereby producing some 17 billion new cells!

Ex vivo studies of rat liver cells show that hepatocyte proliferation is stimulated by chemicals called growth factors. There are many kinds of growth factors. As we have already seen, human growth hormone (hGH) has been used to help children born with the genetic defect of dwarfism to grow to normal size and may be a key weapon against aging.

In a laboratory culture dish, when rat hepatocyte cells are exposed to *hepatocyte growth factor* (HGF) and *epidermal growth factor* (EGF), the cells dedifferentiate, multiply rapidly, then redifferentiate to form adult hepatocytes.

The situation is not that simple, of course. There are many different chemical factors involved, and researchers have not sorted out all of them, nor which are necessary and which may be extraneous. The list includes HGF, EGF, interleukins, TNF-α, norepinephrine, insulin, and others.

---

[20]These functions include glucose regulation, production of many blood proteins (such as albumin and coagulation proteins), secreting bile, breaking down toxic compounds in the blood, and more.

Interleukins are proteins produced by the leukocytes (white blood cells) that promote growth and differentiation. *TNF*-α is tumor necrosis factor alpha, which suppresses tumor growth. *Norepinephrine* stimulates the action of HGF and EGF.

Once the liver has reached its proper size, other chemical triggers come into play and suppress further growth.

While bioscientists work to sort out how the liver regenerates itself naturally, other investigators are looking into the possibility of regrowing severed nerves.

Spinal Cord Injuries

Spinal cord injuries paralyze some fifteen thousand Americans each year. Most famous of these is actor/director Christopher Reeve, injured in a horse-riding accident and now one of the ninety thousand paraplegics and quadriplegics in the United States.

The brain and spinal cord make up the *central nervous system* (CNS). The other nerves of the body are called the *peripheral nervous system*. While most of the peripheral nerves can regrow to some extent after an injury, the nerves of the CNS do not. That is why the effects of Parkinson's disease and Alzheimer's are so devastating. Once neurons in the brain or spinal cord have been damaged, they do not regrow or replace themselves.

Until now.

In 1996 a research group at the Karolinska Institute in Stockholm, Sweden, reported that they had obtained "true functional regeneration" of nerve cells in the severed spinal cords of laboratory rats. At the site of the injury, they spliced peripheral nerves from other parts of the rats' bodies, then used a growth factor to stimulate the nerves' regeneration. Within two months, the rats had regained some use of their hind legs, which had been totally paralyzed.

As Winston Churchill once cautioned: "This is not the end. It is not even the beginning of the end." None of the rats were able to stand or walk. But the basic principle of stimulating nerve regrowth certainly looks promising.

Neurons are composed of three elements: the *dendrites*, which are branching filaments that receive sensations in the form of electrical impulses; the *nerve body*, which processes those signals; and the *axons*, long transmitting fibers that send the processed signals on to other nerve cells (eventually, either to the spinal cord or the brain). Think of the dendrites as sensors, the nerve body as a microprocessor computer, and the axons as telephone lines to other computers.

The Swedish team recognized that spinal neuron axons are coated with *myelin*, a protein that inhibits nerve growth. So once they had severed the rats' spinal cords, they grafted nerve tissue into the gap in such way that it touched the severed nerves' central bodies, not their axons. Then they added *fibroblast growth factor* (FGF), which stimulates growth in peripheral nerves.

In two months the rats stopped dragging their hind limbs. They were able to use the hind limbs to carry some of their weight and even showed some coordination in the movements of front and hind limbs.

This work is all very preliminary, of course. And critics have pointed out that in real cases of spinal injury, the spine is not neatly severed but is crushed and the nerves mangled.

In late 1996 four teams of researchers reported that apoptosis may play a role in preventing such mangled nerves from regrowing. The teams were from Ohio State University College of Medicine, Washington University in St. Louis, Henry Ford Hospital in Detroit, and the Veterans' Administration Medical Center in Kansas City.

Again using lab rats, their studies showed that for three weeks following the injury, nerve cells died apoptotically, apparently triggered into orderly suicide by chemical factors released by the damaged nerves. They found similar evidence in injured monkeys, as well.

By treating the injured rats with *cycloheximide*, a compound that inhibits apoptosis, the researchers found a 50 percent improvement in the animals' ability to use their hind legs after the injury.

Fetal Tissue Transplants

Meanwhile, other researchers are examining the possibilities of regeneration in the brain itself. Between the ages of roughly 20 and 90, the brain actually shrinks, losing about 5 to 10 percent of its original mass. The lost brain mass is replaced by liquid.

Of the 100 billion cells in the brain, only about 10 percent of them are neurons (nerve cells). The rest are *glial cells*, which are responsible for maintaining the correct chemical environment for the neurons so that they can operate at optimum efficiency.

Neurons do not reproduce themselves after birth. The number of neurons you are born with is the maximum number you will ever have. On the other hand, neurons do not age either. They can be injured and destroyed, but cells that do not reproduce do not undergo telomeric shortening and therefore do not age.

The glial cells do age and die. When they die, the neurons they have been protecting are subjected to damage, because the glial cells are no longer there to protect them. In this view, neurons die because their protective glial cells age and die. There is evidence that the loss of glial cells is a factor in diseases such as Parkinson's, Alzheimer's, and many *dementias* (organic brain diseases that cause a loss of mental faculties).

Physical injuries, as well as diseases such as Parkinson's and Alzheimer's, cause damage to the neurons of the brain, leading to loss of memory and physical control of the body. Experiments in which fetal brain tissue is grafted into the damaged area of the brain have shown some promising results. The fetal brain cells are still multipotent and capable of replacing the lost neurons, no matter in which part of the brain they may be.

Transplanted fetal tissue takes on the identity of the cells around it, sends out fresh nerve fibers to other areas of the brain, and makes the connections necessary for that region of the brain to operate properly. As Dr. Kenneth Jon Rose puts it: "Transplanted fetal tissue seems to find its way to the right place in the brain and executes the appropriate physiological function."

But fetal tissue transplant has aroused fear and loathing among antiabortion activists, who apparently fear that such therapy, if successful, will lead women to get themselves pregnant and have abortions so that they can sell their fetuses for clinical use. The controversy reached into the White House, where orders were given to cease funding research that used fetal tissue.

We will examine such fears and political ploys in later chapters.

In the April 4, 1997 issue of *Science*, Ronald McKay, of the National Institute of Neurological Disorders and Stroke, reported that stem cells exist for the central nervous system, "multipotential cells that are the precursors to both neurons and glia." While research in this area has been confined to animals—so far—the implications for regenerating human brain cells are powerful.

For now, we can say that research in regeneration is moving toward the capability of regrowing damaged or diseased organs, such as the heart or kidneys, and repairing crippling damage to the central nervous system.

Research in regenerating skin and allowing wounds to heal without amassing disfiguring scar tissue could ultimately lead to a new era in cosmetic therapies. Even without an elixir of youth, it may be possible to keep the skin smooth and free of wrinkles and spots, while tightening the underlying muscles to produce a taut, trim youthful appearance. Cosmetic surgery may be replaced by regenerative medicine.

Embryonic Stem Cells

The prospects for regeneration took another large leap forward during the summer of 1997.

John D. Gearheart of Johns Hopkins Medical Institutions revealed that his team had isolated the stem cells of a human embryo. Such cells should be totipotent, capable of producing any other type of cell in the human body: blood, nerve, muscle, skin, or any organ.

The embryo's stem cells are the precursors of the development of a complete human baby. They produce all of the body's various 100 trillion cells. If the Johns Hopkins discovery is valid, it should become possible to use embryonic stem cells to regenerate any type of cell that the body needs.

Regeneration will then become not only possible but the preferred therapy for everything from cosmetic repair to replacement of failing organs.

Providing, of course, that the moral, ethical, and political objections to using fetal tissues are overcome. Or will it become possible to clone fetal tissue, thereby removing the need for a continual supply of fetuses? Cloning human tissue, which is also looked upon as anathema by those who object to abortion, may offer a way to produce abundant multipotent or totipotent cells for regeneration procedures.

The future of regeneration research lies just as much in the halls of political power and the pulpits of churches as it does in the laboratories of scientists.

15

Molecular Engineering

There's plenty of room at the bottom.
—RICHARD P. FEYNMAN

∞ "WHAT WOULD HAPPEN IF WE COULD ARRANGE ATOMS, one by one, the way we needed?"

Caltech physicist Richard P. Feynman asked that question in a talk titled "There's Plenty of Room at the Bottom," given at the year-end meeting of the American Physical Society in 1959.

Feynman, who would later win the Nobel prize and write bestselling books about physics, was speculating on the possibilities of working with the smallest units of matter: individual atoms.

Twenty-seven years later a bright young MIT graduate, K. Eric Drexler, gave Feynman's speculation a solid theoretical background and coined a name for it: *nanotechnology.*

Nanomachines

Nano is the scientific/technical prefix for a billionth. It comes from the Latin word for "dwarf." Nanotechnology deals with machines that are measured in *nanometers*, billionths of a meter (a nanometer is not quite a quarter-millionth of an inch). Drexler's envisioned nanomachines are about the size of viruses.

The basic idea behind nanotechnology is that if you could build machines the size of viruses, they would be able to work with individual atoms and molecules. They could build things one atom or molecule at a time. They could take apart things by breaking them up into individual atoms. We would enter an era of molecular engineering, in Drexler's phrase. Nanomachines are molecular manipulators.

Building with Atoms

Think about the way an automobile is built, for example. The process starts by mining ores, which are refined into metals such as steel, aluminum, chromium, copper, and so on. Plastics are manufactured in petrochemical plants out of organic chemicals, mostly from petroleum. Then these materials are shipped to the automobile factory, where they are stamped, molded, welded, or fastened together to produce a finished auto.

This manufacturing process has actually been dealing with atoms of iron, aluminum, carbon, copper, and so on; huge numbers of atoms, atoms by the billions of billions. This is what Drexler calls *bulk processing*. It is the way we have built our tools and gadgets since time immemorial.

Nanotechnology offers a better way, Drexler believes. Nanomachines could take a pot of ingredients, one atom or molecule at a time, and build your automobile by placing those atoms and molecules just where they ought to go. Nanomachines could link atoms together precisely into the exact molecular structures that you desire. Building an auto would be more like a swarm of invisible genies working silently and swiftly than a noisy, clanking factory.

Moreover, nanomachines could build automobiles out of much better materials than today's metals and plastics. Take a pot of charcoal dust—ordinary carbon soot—and nanomachines can produce out of it an automobile of pure diamond. By arranging the carbon atoms precisely, they could build a car with the structural strength and lightness of diamond—at a fraction of the cost of today's metal and plastic autos.

175

Now picture nanomachines at work inside your body, breaking down plaque on your artery walls, tearing apart invading microbes (such as HIV) like an extra and super efficient immune system, delivering molecules of medicinal drugs precisely to the cells where they are needed, destroying tumors like a nanometer-sized surgical team, rebuilding worn or diseased tissues wherever they are found, molecule by molecule, atom by atom.

It sounds too good to be true, and the fact is that nanomachines exist today only in the dreams of farsighted engineers and scientists. But Drexler's vision is more than a dream. Research on nanotechnology is under way at universities and laboratories in Britain, Japan, and elsewhere, as well as MIT, Stanford, Cornell, the University of Michigan, IBM, Lucent Technologies' Bell Labs, SRI International, and other research institutions in the United States.

To date, the most publicized success of nanotechnology has been accomplished by physicist Don Eigler at the IBM research laboratory in Almaden, California. He used a scanning tunneling microscope to move individual atoms of xenon across a surface of nickel to spell out the initials IBM in letters that were only five nanometers high.

A scanning tunneling microscope uses an electrical field to nudge individual atoms around with nanoscale precision. But although the microscope can manipulate atoms and make their positions visible, it is attached to a large roomful of equipment. That is a long way from the independent, virus-sized nanomachines that Drexler envisions.

Assemblers and Dissociators

Drexler sees nanomachines that are mobile and armed (literally) with grasping devices that can grab individual atoms. The kind of nanomachines he calls *assemblers* will be directed to take atoms (or perhaps molecules, groups of atoms) and put them together—assembling something such as a flea-sized computer or a full-sized four-door sedan.

Nanomachines could also be used as *dissociators* to take something apart, atom by atom. Cholesterol plaque builds up along artery walls and narrows the arteries' inner diameters, restricting blood flow and leading to atherosclerosis. Dissociators could be programmed to search out plaques and take them apart, breaking them down into small molecules or even individual atoms that can be carried off by the bloodstream to the kidneys for elimination from the body.

Big Bugs . . .

How do you build a nanomachine?

Feynman proposed a sort of evolutionary approach. He thought the way to nanomachines was to build smaller and smaller devices. Possibly he was thinking of the old saw:

> *Big bugs have little bugs*
> *Upon their backs to bite 'em.*
> *Little bugs have littler bugs,*
> *And so, ad infinitum.*

In 1959 Feynman offered a prize of $1,000 to the first person who could shrink the information on a page of a book to one twenty-five-thousandth its original size, and keep the letters legible enough to be read by an electron microscope. Feynman's sense of humor must have come into play here: Reducing a page by a factor of twenty-five thousand would be enough to put the entire Encyclopaedia Britannica on the head of a pin.

He also offered another $1,000 prize to the first person to make a working electric motor no larger than a cube one sixty-fourth of an inch on a side—about the size of the period at the end of this sentence.

Most of Feynman's audience thought he was joking. Yet within a few years he paid off both prizes.

Drexler agrees with this evolutionary approach and envisions a period of development in which engineers build the smallest machines they can, and these machines in turn build still smaller units, until a nanometer-sized *replicator* has been perfected. As its name suggests, the replicator would be designed to build working nanomachines.

The Limits to Reduction

Is it possible to make machines the size of viruses? Machines that will work as programmed?

Not today. Critics of nanotechnology say it will never be possible. The world of the ultrasmall is very different from the macro-sized world to which we are accustomed—and adapted.

An object that is only a few nanometers across is buffeted by thermal vibration, the motions caused by heat energy. Down at that scale, atoms and molecules are constantly jiggling; nothing stays still. The hotter the temperature, the more thermal vibration there is. In fact, the vibrations of the particles *are* heat.

When you measure your body temperature, for example, the thermometer's mercury rises because the molecules of your body are banging against the thermometer and transferring their thermal energy to the mercury. The rest of the thermometer is heated too, but the mercury is a liquid and therefore expands much more easily than the metal and glass parts of the thermometer.

178

Biochemists have seen one manifestation of thermal vibration in their microscopes since 1827: Brownian motion, the jiggling of microscopic-sized particles, such as dust grains, caused by the constant jostling of the molecules in which the particles exist.

How can you make a nanomachine do any useful work, the critics ask, when it will be constantly bombarded by the thermal vibrations of the molecules around it? Build structures by placing individual atoms precisely where they are wanted? Impossible, claim the naysayers. It would be like trying to do brain surgery in a jam-packed, lurching, speeding subway train.

Worse yet is the factor of *quantum uncertainty*. Physicists have learned that the ultrasmall world of the atom does not behave in the familiar way of our macro-world. Werner Heisenberg showed, in 1927, that it is impossible to know both the position of an atom and its momentum (the direction and speed of its motion) at the same time.

Quantum effects, say the critics, will make it impossible to guide or direct nanomachines even if you could build them.

To this criticism, Drexler and his colleagues point out that nanoscale devices already exist and seem to work perfectly fine. Viruses move molecules about with considerable precision, despite the buffetings of thermal vibration and the effects of quantum uncertainty. If viruses can work, why not virus-sized machines?

Only further research and experimentation will prove which side of this argument is correct. In the meantime, though, progress is being made.

Nanobatteries and Brownian Energy

There are other questions about nanomachines besides. What will be their energy source? How do you power a virus-sized machine? Small or not, it must have some source of energy.

And how do you direct them? How are they controlled? If a swarm of nanomachines are let loose inside your body to clear your arteries of plaque, how do you program them to do that? And how do you stop them from going on to tear down your artery walls, instead of the plaque?

Both the friends and foes of nanotechnology worry about "the gray goo problem": that is, the possibility that nanomachines might inadvertently start chewing up the landscape, uncontrollably destroying everything in their path.

The energy problems, at least, are already being addressed.

In 1992 Reginald Penner and coworkers at the University of California, Irvine, produced a battery that is only seventy nanometers across, about the size of an adenovirus. Built of silver and copper, it put out twenty thousandths of a volt for forty-five minutes. That does not sound like much, but remember that nano-sized machinery will not need a lot of power.

The battery is far from being a practical device, but it shows that nanoscale energy systems are certainly possible.

In 1997 a group of biochemists suggested using Brownian motion to power nanomachines. Brownian motion, as we have seen, is the result of thermal vibrations that were claimed to be a barrier to nanotechnology.

It may be possible, according to R. Dean Astumian and his colleagues at the University of Chicago, to use chemical energy to create a preferred direction to the pushes of Brownian particles, a sort of chemical equivalent of a ratchet wheel that turns in only one direction.

Drexler and others have proposed using the body's internal heat as the energy source for nanomachines inserted into humans. Programming and directing those nanomachines, they believe, are problems that will be solved. One possible road to a solution is to make each set of nanomachines sensitive to only one form of molecule—a "solution" that the lymphocytes of the immune system worked out hundreds of millions of years ago.

The Size of Things to Come

While full-fledged nanomachines are still some distance in the future, research in nanotechnology and allied fields will pay handsome dividends much sooner.

Bell Laboratories researchers have produced a nanotransistor only 60 nanometers across—a mere 180 atoms wide. It consumes only about a hundredth of the electrical power that ordinary transistors need, yet produces higher current flow and offers the possibility of much higher switching speeds than conventional transistors.

However, quantum effects may make it difficult or even impossible to build still-smaller nanodevices. Already, quantum effects have been detected in the nanotransistor's insulating layer, which is only three atoms thick.

Meanwhile, electrically conducting wires only a hundred atoms across have been built and operated. They could be used to power ultrathin catheters that carry microminiaturized cameras for examining the body's interior the way spacecraft cameras send images from alien planets. They could also power extremely small lasers that might one day be used as surgical scalpels within the body with no more damage to the skin that a pinprick. Major incisions—and the scars they leave—may become a thing of the past.

MIT engineers A. H. Epstein and S. D. Senturia have shown that a gas turbine generator of about one cubic centimeter in volume could, with today's technology, produce fifty watts of electrical power or, if used as a jet engine, about four hundredths of a pound of thrust. Not much for a jet airliner, but a stack of such microthrusters could easily propel a miniature drone aircraft laden with sensors—or a miniature roving vehicle on the surface of Mars.

They point out that such miniaturized gas turbines can deliver ten to thirty times more energy, pound for pound, than even the most advanced batteries. A millimeter-sized engine would be ten times more efficient than a modern jet aircraft engine.

Thus there are solid incentives to drive toward smaller mechanisms. Micromotors, miniature heat exchangers, even miniature chemical reaction chambers have been built. While all these precursors to nanomachinery deal with relatively low-power applications, the medical uses of nanotechnology will not require high powers.

Nanotherapies

One day disorders as diverse as obesity and cancer may be treated by swallowing a glass of orange juice—which contains hundreds of millions of nanomachines. They will spread through the body, a sort of auxiliary immune system directed by human intelligence (and they will have to either avoid or negate attack by the body's natural immune system).

These nanomachines will be bodyguards that can protect against infection, rebuild worn muscles and organs, strengthen failing bones, keep the arteries clear of obstructions, and perhaps even enhance natural intelligence.

It sounds like science fiction, of course. But so did organ transplants and gene therapy, not so long ago.

Remember the slogan of *Amazing Stories*, the first magazine devoted entirely to science fiction: "Extravagant fiction today; cold fact tomorrow."

The "tomorrow" for Drexler's vision of nanomachines may be many years in the future. But the time will come.

182

PART TWO

THE IMPACT OF IMMORTALITY

∞

Arthur C. Clarke once wrote that every revolutionary idea evokes three stages of reactions:

1. At first people say, "It's completely impossible."

2. Then they say, "Maybe it's possible to do it, but it would cost too much."

3. Finally they say, "I always thought it was a good idea."

In the pages to follow, we will examine how the prospect of human immortality will affect our lives, our society, and the entire human race. And we will meet each of Clarke's three reactions.

16

"It's Completely Impossible"

The history of molecular biology reminds us how quickly the unimaginable is translated into routine.
—PHILIP KITCHER

∞ WHEN I WAS A TEENAGER, FLYING TO THE MOON WAS regarded as the ultimate impossibility.

I remember hearing in 1948: "Harry Truman has as much of a chance of getting reelected President as we have of flying to the Moon." Truman won an upset victory that November. A little less than twenty-one years later, Neil Armstrong and Buzz Aldrin set foot on the Sea of Tranquility.

Nuclear power, organ transplants, desktop electronic computers, supersonic aircraft, coronary artery bypass operations—all were once regarded as totally impossible. Yet, over the years of my own lifetime, Buck Rogers's ray guns became lasers, astronauts flying into space have become so routine that the news media only covers space flight when there is an accident or controversy, and the "old folks" have become the "golden agers" who play golf and tennis, take ocean cruises, and power the economy of states such as Florida and Arizona.

Optimists and Pessimists

The pessimist sees the glass half-empty, the optimist sees it half full. Let me tell you why I am an optimist.

I started my working career as a reporter on a weekly newspaper in suburban Philadelphia. During the summer, we carried a box score on the front page every week—about polio: how many children had been killed by the disease the previous week; how many were crippled; how many had to be placed in iron lungs so that they could breathe.

One fine spring day we carried a story about kids getting Salk vaccine shots. Great human interest photographs of children grimacing and wincing as doctors jabbed needles in their arms while their mothers stood bravely tearful in the background.

And we never had to run a polio box score again.

I moved from newspapering into the aerospace industry to work on the first American space project, *Vanguard*. In those early days of the so-called "space race" between the United States and the Soviet Union, the Russians scored most of the successes and our rockets tended to blow up. The first attempt to launch a *Vanguard* satellite ended in an ignominious and quite spectacular explosion a few feet above the launch pad on December 6, 1957.

Learned mathematicians published articles showing, with impeccable logic, that rockets are such complicated machines that it was statistically impossible to expect each and every part of a rocket launcher to work correctly and in sequence. That was why the rockets blew up. They would always blow up, the mathematicians asserted.

The fact that Soviet rockets seemed somehow to evade this mathematical certainty did not enter into their calculations. They were wrong, of course. Solid engineering and hard-earned experience overcame the shortcomings of our early rockets. We went on to the Moon and have sent robot probes to all the planets of the solar system, save distant-most Pluto.

Today when I tell friends and colleagues that human immortality is in sight, they stare at me in disbelief.

It's impossible, of course. You might as well try flying to the Moon.

The preceding pages of this book have shown why I believe that human immortality is not only possible but near. Men and women alive today may well be able to live for centuries, and if they survive that long, they will undoubtedly be able to live for millennia. We are reaching the point in our knowledge of biology and medicine where death from aging will no longer be inevitable.

When I discuss the matter rationally with friends and colleagues, what comes out is not that they regard human immortality as scientifically impossible. They grudgingly admit that if you give enough scientists enough time (and funding), they can accomplish almost anything.

No, they do not really believe immortality is impossible.

They believe it is undesirable.

They are not saying, "It can't be done." Their true reaction is: "It *shouldn't* be done."

Which brings us to a sheep named Dolly.

Questions About Cloning

In February 1997 a team of scientists from the Roslin Institute in Scotland rocked the world by announcing that they had successfully cloned a sheep.

Dolly was produced not from an egg fertilized by a sperm cell but by taking a sheep's egg cell and replacing its DNA-containing nucleus with the nucleus of a cell from a 6-year-old ewe's body. The egg cell was then placed in another ewe's womb and developed normally.

The lamb, named Dolly, was genetically identical to the sheep from which the cell nucleus had been taken.

No one had cloned a mammal from an adult cell before. But cloning was not entirely new. It had been going on since the early 1980s, actually. That is when biologists developed a procedure in which they replaced the nucleus of an egg cell with a nucleus from another egg cell. Monkeys were cloned that way.

But the altered egg could develop into a clone of the original animal only if the replacement nucleus came from a cell of a barely developed embryo. Cloning attempts using nuclei from adult animals invariably failed.

In 1996 the Scottish team, led by Ian Wilmut and Keith H. S. Campbell, successfully cloned sheep from older embryonic cells by placing the intended donor cells in a nutrient-deprived medium, essentially starving them for five days. This forced the cells out of their normal growth cycle and into a quiescent stage. For reasons still under study, nuclei from these cells are more readily accepted by eggs.

Dolly was the first mammal to be born not from an embryo's cell but from a somatic cell of an adult. Theoretically, this means that it should be possible to clone animals, eventually including human beings, from any kind of donor cell.

Although many biologists feared that the DNA inside the nuclei of adult cells undergoes irreversible changes as the cells differentiate and specialize, Dolly's birth shows that the DNA in an adult nucleus either reprograms itself or is open to reprogramming by factors in the egg.

The question is: How old is Dolly?

"Our seven-month-old lamb actually has a six-year, seven-month-old nucleus in all her cells," said Grahame Bulfield, director of the Roslin Institute. "It's going to be interesting to see what happens with the aging of this animal."

Does the cell's DNA harbor a clock that determines the organism's physiological age? If so, is that clock the telomeres that cap each chromosome or some other mechanism among the genes as yet undiscovered? And is the clock somehow reset by the act of cloning?

The answers to those questions are of primary importance in extending human life span.

Another critically important question is: Will it be possible to add or delete genes from the donor DNA before producing clones from it? In other words, will bioscientists one day be able to clone organisms with DNA tailored to suit a preconceived demand? Do you want your cloned sheep to produce more wool than normal sheep? Do you want to clone a racehorse that can outrun any thoroughbred?

189

The answers will not be long in coming. In July 1997 the Scottish team announced the birth of five more lambs, all cloned from fetal cells. But these lambs carry extra genes; some even have a human gene that was inserted into the fetal cells before they were cloned.

This is part of the effort undertaken by PPL Therapeutics, a Scottish biogenetics firm that worked with the Roslin Institute team, which is reportedly working on replacing the genes that code for blood plasma in sheep and cows with human plasma genes; thus the animals' milk will contain human blood plasma elements, including albumin, clotting factors, and antibodies.

Genetically altered animals could produce thousands of times more human blood plasma per year than human blood donors now give. By the end of 1997 the Roslin Institute scientists announced the birth of Molly and Polly, two more cloned lambs who have been genetically engineered so their milk will contain blood-clotting factors that could be used for treating leukemia in humans.

The ultimate question is, of course: Will it be possible to clone humans? And if so, do you want to clone copies of yourself? Should they be exact copies or perhaps a little taller and slimmer? And have blue eyes instead of brown?

The Reaction to Cloning

No sooner was Dolly's birth announced in Edinburgh than a worldwide cry of fear and revulsion issued from the throats (and keyboards) of politicians, religious leaders, and media pundits.

The President of France, Jacques Chirac, called cloning "a degrading attack on humanity" and suggested a ban on such research throughout the industrialized world.

Pope John Paul II also inveighed against human cloning, fearing that scientists will attempt to "play God." Later the Vatican announced that cloned human beings would not have souls, a point of view that many dreadful sci-fi movies have been playing on for generations.

U.S. President Bill Clinton directed the National Bioethics Advisory Commission to prepare a report examining the ramifications of cloning technology and urged that no federal funding should be allowed for research into cloning human beings.

Later in 1997 the National Bioethics Advisory Commission recommended that the moratorium on human cloning research be continued and reinforced with federal legislation, provided such laws have a "sunset clause" that will allow for a re-examination of the situation in a few years and are written carefully enough so that other forms of cloning research are not outlawed.

The general reaction pounced on the idea of cloning humans, of "playing God." But why? Why did media commentators, politicians, and religious leaders *immediately* react with such expressions of fear and outrage?

It is important to understand this reaction because this kind of knee-jerk negativism is what we can expect if and when the general public realizes that human life spans can be extended for centuries or more. The public reaction to cloning is a nearly perfect example of what the reaction will be to the prospect of human immortality.

This reaction is puzzling and troubling, because a cloned human being will be just like any other human. The Vatican may believe that a cloned person does not have a soul, but souls are ephemeral things, not measurable by scientific methods. Besides, if God creates souls for people born the normal way, why not for clones?

This hostile opposition of the naysayers betrays a fundamental ignorance of what cloning can and cannot do.

Yes, cloning can produce an offspring that is *physically* an exact copy of its parent—if successful. At present, even cloning a sheep took some three hundred attempts before Dolly was produced.

But a cloned human being will be no more a slave, a soulless zombie, or an exact copy of its parent's *personality* than any two identical twins are zombies or have exactly the same personalities. The sum total of a human personality depends not merely on his or her genes, but on the experiences the person has throughout life.

Remember the difference between genotype and phenotype that was discussed in Chapter 9. A person's genes, by themselves, do not determine that person's fate. A cloned egg is still subject during gestation to the specific chemical factors of its mother's womb, factors that are unique to that woman and that pregnancy. Genetic determinism is no more valid than the belief that the Earth is flat or the phlogiston theory of heat.

Meanwhile, Back at the Lab . . .

Look at the other side of the coin. Yes, there may be fears and dangers in cloning research; every new area of knowledge is filled with unknown possibilities.

But what happens if we stop research into cloning?

One of the motivations of the Scottish work was the hope that it may become possible to create animals such as sheep or cows that can be genetically altered to produce pharmaceutical chemicals—medicines—in their milk. As we have seen, genetically altered animals could produce human blood plasma and then be cloned to meet the need for plasma worldwide.

All right, perhaps animals can be developed into biochemical factories. Punsters are already talking about "pharm" animals. An animal breeding firm in Wisconsin, ABS Global, revealed in August 1997 a 6-month-old cloned calf, Gene. ABS Global has spun off a new company, Infigen, specifically to clone transgenic cows capable of producing a variety of therapeutic proteins in their milk.

But what about cloning humans? Isn't the old-fashioned way of producing babies good enough?

Yes, but . . .

Some babies are born with defects in the mitochondria of their cells, those DNA-bearing power plants that produce the cell's energy from nutrients and oxygen. Mitochondria with defective DNA can cause many types of devastating illnesses, including blindness.

If it becomes possible to remove the defective mitochondrial DNA and replace it with healthy DNA cloned from a donor, such diseases can be eliminated.

Organ transplants often fail because of tissue rejection; the body's immune system attacks the transplanted organ because it recognizes the tissue as "foreign." If it becomes possible to clone a new heart from the patient's own tissue, the transplant procedure becomes infinitely easier. If it becomes possible to clone healthy brain cells, the ravages of Parkinson's, Alzheimer's, and even traumatic brain injury can be reversed.[21]

Stop cloning research and these opportunities will be lost.

In the meantime, ongoing research brought up a new possibility. The 1997 announcement from Johns Hopkins that researchers had discovered the embryonic stem cells that give rise to all the other cells of the developing embryo may well replace cloning as a political "hot potato."

Such totipotent cells could be used to generate blood, nerve, muscle, or the cells of any organ. This discovery may make organ and tissue regeneration practical. But since they are derived from human embryos—aborted fetuses—much the same furor can be expected as any other therapy touching on the explosive abortion issue.

"Playing God"

People tend to fear new possibilities. And the easiest thing to do with something that frightens you is to banish it. Get rid of it! Then it can't hurt us. Yes, but it can't help you either.

In attacking the idea of cloning, religious leaders and media pundits trotted out one of their favorite catch-phrases: "playing God."

When Pope John Paul II invoked this cliché, His Holiness overlooked the fact that the medical attention that has more than once saved his own life is surely "playing God." Otherwise, he would be among the angels now.

---

[21]At present, research has shown that transplanted fetal brain tissue can help repair such damage, but the use of fetal tissue has stirred the wrath of those opposed to abortion and thus has been slowed or halted altogether by political decisions in the United States and elsewhere.

Every time we take an aspirin or an antibiotic, we are "playing God." What else is the coronary bypass procedure that saves a heart attack victim from imminent death?

When anesthetics were first introduced in surgery in the early nineteenth century, moralists railed against their use in childbirth. The Bible commands that women bear children in pain and suffering, they thundered, therefore giving the laboring mother something to ease her pain was against the dictates of God's expressed will.

But when Queen Victoria, no wild-eyed radical, decided that *she* would use anesthesia in childbirth (she had nine children), the moralists were silenced. A blow had been struck for the advance of medicine and science. And feminism.

People fear new ideas. Resuscitating a patient whose heart has stopped may be "playing God" the first few times it is done; afterward it becomes a standard part of emergency medical treatment, and we take it for granted.

One of the reasons for the fear of cloning is that many people associate the idea of cloning with the concept of trying to produce some sort of "ideal" human beings through eugenics, by deliberately selecting genetic traits such as height, eye color, and so on. Especially to those who remember Nazi Germany's misguided and murderous quest for a "master race," eugenics is anathema.

Yet it would be the utmost folly to throw away the possible benefits of cloning over fears of its misuse. A eugenics program dictated by a government, where only certain physical types are regarded as desirable (or even permissible), is certainly a form of tyranny that should be resisted by all. But an individual's desire to produce offspring that are as close to the ideal that the parent can envision seems well within the rights of any citizen.

We will face the same objections and emotional reactions when human immortality becomes a public issue.

These objections boil down to a choice between pessimism and optimism, a struggle between hope and fear.

The pessimists fear that the new knowledge will be used in harmful ways; they are willing to forego the possible benefits for fear of the possible harm.

The optimists look forward to the possible benefits from new knowledge, and believe (perhaps naively) that the possible harm can be avoided, minimized, or controlled.

195

Death with Dignity

Today more public attention is focused on how we die than on the prospects of avoiding death altogether.

Modern biomedical technology allows terminally ill patients to be sustained long past the point where their bodies could continue living without such help. Many people fear that their lives will be unnecessarily prolonged—and their estates drained to the last cent—by such "heroic" methods.

People are demanding the right to die with dignity, rather than being maintained in a vegetative state by life-support machinery. Some demand the right, when faced with a hopeless medical condition and the prospect of physical agony and emotional anguish, to decide to end their own lives through doctor-assisted suicide. Dr. Jack Kevorkian has become famous (or infamous, depending on your point of view) for defying local laws and assisting patients to die with as much dignity and as little pain as possible.

The Supreme Court decided in 1997 that the U.S. Constitution does not include the right to commit suicide. The separate states will have to decide if they want to legalize suicide and enact their own laws on the subject.

As we have seen, a few people have already decided

to have their bodies frozen after being declared clinically dead in the hope of one day being thawed and cured of the ailment that "killed" them. Instead of suicide, they look to a life in the future.

My father did not die with dignity. Even when taken off life support, he struggled on in a coma for weeks, unseeing, unhearing, but desperately fighting to keep on breathing to stay alive. He died at last when he had no strength left to continue the battle.

Do I want to die like that? Or would it be better to go quietly when I become certain I cannot live much longer?

If the trends of biomedical research are leading toward immortality or, at least, huge extensions of human life span, do I want to go at all?

Who Wants to Live Forever?

Opponents of gun control insist that: "When guns are outlawed, only outlaws will have guns."

Opponents of immortality—or vastly extended human life spans—have already voiced the opinion that only megalomaniacs will want to live forever. Think of an immortal Hitler or Stalin or Genghis Khan.

Do *you* want to live forever? Is it morally right to try to extend our life spans past the traditional three or fourscore of years?

We are dealing with life and death quite literally here. The ultimate horror, the transcendent fear, death in all its finality may be avoidable. The prospect threatens our most fundamental beliefs, shatters our basic understanding of life and its fragility (see Pandora and the Origin of Death, opposite).

Moreover, by taking away the fear of inevitable death, we may be taking away the most compelling support for religion. It is that fear of death, what Shakespeare's Ham-

PANDORA AND THE ORIGIN OF DEATH

Every culture has a myth about how death entered the world. In most of them, death did not exist when the first mortals were created. An act of disobedience by the first humans angered the Creator, who punished the humans by making them mortal and susceptible to death.

This is the essence of the Biblical story of Adam and Eve in the Garden of Eden.

Similarly, the ancient Greeks had their legend of Pandora, the first woman, created by Athena and Hephaestus. Men had been created earlier by Zeus.

Pandora was given a sealed casket by the gods and was told that she must not open it. Curiosity overcame Pandora and her lover, Epimetheus. They opened the casket (as the gods had expected) and out flew all the evils and bad fortune that humankind has been afflicted with ever since: plague, pestilence, strife, woe, sorrow, grief—and death.

Realizing what a foolish and horrible thing they had done, Pandora and Epimetheus slammed down the casket's lid. It was too late to keep all the evils from escaping, but they trapped hope—which had been at the bottom of the casket—inside.

That is why death reigns among us today, the ancient Greeks believed. And why hope whispers to us, muffled inside Pandora's casket, unable to speak louder than a murmur.

let called "the undiscover'd country, from whose bourn no traveler returns," that started the earliest humans on their quest for a life beyond the grave.

Religious thinkers have created the most important

moral codes that we possess. Our society is based on the bedrock of Judeo-Christian morality. Yet if men and women can achieve immortality *here*, on Earth, the most basic motivation for following the ethical teachings of religion is greatly weakened—perhaps fatally weakened.

Thus religious thinkers may have ample grounds for resisting the opportunity of achieving human immortality.

I believe there is no moral injunction against our efforts to extend life span, just as there was no moral injunction against the practice of medicine. Perhaps I won't want to live forever, but it's a choice I would like to be able to make for myself.

In essence, we all make that choice when we go to the doctor for help, when we seek a cure for our ailments, even when we eat properly and exercise. We are trying to prolong our lives. We don't cross busy streets with our eyes closed. Rational people try to avoid death, unless they are so ill (physically or mentally) that they see death as preferable to life.

Maybe, if I get to be 90, or 900, or 9,000, I will decide that I want to end it all. Perhaps, like the suave movie actor George Sanders, I will take my own life and leave a note that simply reads: "I am bored." But I would prefer to have the option in my own hands and not leave it to the blind workings of chance. Or free radicals. Or glucose browning. Or telomere shortening.

And once it becomes known that human life span can be greatly extended, the moral questions will be swept aside in the stampede to get whatever it takes to keep on living. As Austad puts it:

> It does seem compelling apparent, though, that regardless of the social desirability of slowing aging, if science uncovers therapies that can do it, those therapies will be employed. This is one genie that has no chance of being put back in the bottle.

This is not selfishness, or greed, or any other manifestation of moral degradation. It is nothing less than what we should expect. Life seeks life. Living organisms strive to continue living. They do not ordinarily go gentle into that good night.

But if people start living for centuries, what happens to society? How can the world hold together if (and when) humans stop dying of old age?

17

"It Would Cost Too Much"

It is the preservation of the species, not of individuals, which appears to be the design of Deity throughout the whole of nature.

— MARY WOLLSTONECRAFT GODWIN

∞ ONE OF THE FIRST FEMINISTS, MARY WOLLSTONECRAFT Godwin, died eleven days after giving birth to the daughter who would become Mary Wollstonecraft Shelley, author of *Frankenstein*.

Mrs. Godwin observed that the forces of nature seem more concerned with preserving species of organisms rather than preserving individuals. This appears to be confirmed by what we have learned in the earlier chapters of this book; our bodies begin to age shortly before puberty, as if they are designed to last only long enough to ensure that we produce offspring.

It is not that straightforward, of course. From that strictly utilitarian viewpoint, what's the sense of having us live into our sixties, seventies, and longer? Once we have produced children and nurtured them to the point where they can have children, we become expendable, as far as the forces of nature are concerned. Grandparents are modern luxuries, where *modern* means within the past few thousand years.

Yet a biologist will maintain that evolution does not happen on the level of a whole species. Evolutionary forces, those blind workings of nature, happen to individual organisms. It is the individual who must find food and avoid becoming somebody else's food. The individual must seek out a mate and reproduce and then shelter and nurture its young. It is each individual member of the human species who struggles to learn, to grow, to survive. It is the sum of the fates of all the individuals in a species that determines whether the species survives or vanishes into extinction.

Societies follow similar biological imperatives. Groups of human beings, from cave-dwelling clans of a dozen people to modern nations of hundreds of millions, also struggle to survive. Societies can change much faster than our physical bodies change; after all, our bodies are still the same as they were in the Neolithic time of hunting and gathering, but our society has changed enormously. Ask any Russian.

Yet societies, like organisms, resist change. This is as natural as life itself. Your body is protected by an immune system that recognizes and kills invading microbes and infected cells. The immune system also attacks cells of transplanted organs, mistaking these foreign cells for dangerous invaders. Societies too have defense mechanisms against "foreign" ideas and forces.

In earlier times we burned people at the stake or stoned them when they espoused ideas or lifestyles that society did not like or trust. Today unpopular ideas are attacked by religious and political leaders and by commentators in the mass media. The brutality of those earlier years has been eased, but the basis of the attack against new ideas seems much the same: "We've never done that before, therefore it must be wrong. Or maybe evil."

A startling new idea, such as human immortality, triggers these societal defense mechanisms. Moralists and religious leaders point out that since aging and death have always been with us, it must be wrong to try to avoid them. Or maybe evil. Politicians will nod their heads in agreement and then ask a couple of more practical questions:

How much is this going to cost?

Who's going to pay for it?

202

The Graying of Society

In the industrialized nations of the world, average life expectancy has doubled over the past century. There are more old people in those societies, and they are more vigorous and active than the elderly of earlier generations ever were. Witness the growth of the sports, entertainment, and travel industries over the past two decades.

Today there are more retirees in the United States than teenagers. This has given rise to powerful organizations, such as the American Association of Retired Persons (AARP), and less formal groupings, such as the Gray Panthers. The elderly make up a powerful political lobby in Washington and the fifty state capitals.

To date, their political activities have been mainly devoted to matters of government policy concerning the financial and medical problems of the aged. They are very concerned about Social Security and Medicare, for example.

As society becomes grayer, these government programs become more and more desperate for funds. Who thought, in 1935, that Social Security would have to pay retirement benefits to tens of millions of men and women who live well into their seventies and eighties? Average life expectancy in 1935 was barely 60 years. Medicare, the government-backed medical insurance program for the elderly, is also under growing financial strain.

These programs are funded by tax dollars. Social Security was budgeted at $369.4 billion in fiscal year 1997, Medicare at $189.9 billion. By itself, Social Security is the largest single item in the federal budget, nearly 50 percent bigger than the budget of the Department of Defense.

The growing number of retired persons forms a political bloc of increasing power that demands better benefits and lower tax rates for themselves. Younger taxpayers, however, are loath to pay out ever-increasing shares of their incomes to support the elderly.

This situation was illustrated cleverly by cartoonist Michael Ramirez in the July 28, 1997 issue of *USA Today*. His cartoon showed an elderly man labeled AARP holding a tin cup labeled MEDICARE and confronting a schoolchild with the demand: "All right, kid, hand over your lunch money."

Currently, the expenditures for Medicare are rising about 8 percent per year, partly because Medicare recipients are living longer than anticipated when the program was first put into action in 1965. In 1997 Congress and the White House struggled over competing plans for "fixing" Medicare. At issue were suggestions to raise the eligibility age from 65 to 67, to have persons earning more than $50,000 per year pay more for their benefits than those earning less than $50,000, and to charge a nominal fee for home visits by health-care givers. Those provisions were defeated in 1997, but they will inevitably rise again.

It seems obvious that such steps, which are the subject of intense political wrangling, are little more than a finger in the dike. On the other side of the dike is the flood of "elderly" who will live for centuries. Or more.

How do we pay for it?

This is a question that must be faced now, today, not in some unspecified future. It is not a problem that can be shelved for some later date.

The reason why is because research on extending the

human life span needs to be funded. As politicians and social leaders begin to realize where this research is heading, the easiest way to avoid the problems that immortality will bring is to throttle down today's funding for such research. Or stop it altogether.

We have seen the immediate response to the cloning of a sheep. "Stop this research!" is the cry. "Don't let them go any farther!"

In the very near future, the same voices may well begin crying, "Stop research on life extension! Don't let them go any farther! They're trying to play God!"

They might as well be saying, "There are some things that Man was not meant to know, Dr. Frankenstein."

Beneath their cries will be the fear that life span extension will totally bankrupt Social Security, Medicare, and the entire private health and life insurance industries.

Who Pays for Research?

Just as "the hand that rocks the cradle is the hand that rules the world," the hand that funds biomedical research is the hand that determines how much we will learn about biology and medicine.

Organized scientific research began in the late sixteenth and early seventeenth century times of Galileo and Kepler. In those days, research was done principally by men wealthy enough to pay their own way or by men smart enough to attract a rich patron to support them. Galileo spent a considerable amount of his time wheedling potential patrons for positions, a chore scientists have had to deal with, one way or another, ever since.

Isaac Newton was the son of a well-off Lincolnshire landowner. He worked out the laws of motion and gravitation while at his parents' country home to escape the bubonic plague that was ravaging Cambridge and London

204

in 1665. Michael Faraday, who did pioneering work in electricity in the early nineteenth century, was a bootmaker's son and started his scientific career as a lab assistant— little more than a janitor.

Science, like most professions, was almost exclusively a male prerogative; it was not until the late nineteenth century that women won enough civil freedom to enter scientific fields, and even then they had to overcome enormous chauvinism and outright hostility.

205

Physicians often did medical research when their practices afforded them the time and money to do so. William Harvey, physician to British monarchs James I and Charles I, discovered the circulation of the blood in the early seventeenth century. Alexander Fleming was a physician who became interested in finding ways to combat bacterial infections when he saw wounded soldiers die of such infections in World War I; in 1928 he discovered penicillin, almost by accident.

Universities were the major sources of research funding prior to World War II; their monies came from wealthy graduates and donors who contributed grants. Philanthropy became institutionalized in the twentieth century, with organizations such as the Carnegie Foundation and the General Education Board (founded by John D. Rockefeller) donating hundreds of millions of dollars to medical schools and universities in the United States.

National governments poured massive funds into research during World War II, harnessing scientists and engineers to produce not only weapons, such as radar and nuclear bombs, but medical breakthroughs, such as antibiotics, as well.

After the war, the U.S. federal government became the major source of funding for virtually every field of research, from astronomy to zoology. The National Science Foundation (NSF) was created in 1950 to fund basic reotsearch in nonmedical fields.[22] The National Institutes of Health (NIH) concentrates on supporting medical and biological studies.

Today the American taxpayer supports close to $75 billion of research and development per year, ranging from NASA's exploration of outer space to the Human Genome Project. Of that amount, slightly less than $15 billion goes to basic research. NIH was budgeted for $12.7 billion in fiscal year 1997; NSF $3.2 billion. (Their budgets include items that are not considered basic research.)

In the biomedical field, most research grants are relatively small, under $100,000 per year. They pay for an individual scientist and his or her assistants and equipment. Scientists do not get rich on government grants. Indeed, wealthy scientists are very rare, and those that are well-off either were born with money or "spun off" their research results into a profitable commercial venture.

There is a great deal of profit—and money invested—in commercial biomedical companies. The pharmaceutical industry does hundreds of billions of dollars in sales each year. Commercial and legal battles are now being fought over the patent rights to genetically engineered organisms, ranging from bacteria to mice, that are worth billions. The U.S. Patent and Trademark Office is squabbling with the National Academy of Sciences over whether or not the government has the right to offer patents on fragments of human genes.

Most of the basic research in the biomedical fields, however, is funded in the United States by tax dollars. Much less than a penny out of every tax dollar is spent on biomedical research. Yet control of the tax purse strings

IMMORTALITY

206

---

[22]It is sobering to find that neither the Encyclopaedia Britannica nor various "timetables of history" books nor David McCullough's magisterial biography of President Harry S. Truman even mentions the National Science Foundation.

determines, to a large extent, which research programs are carried out and how quickly they can be done. Progress in using fetal brain tissue to repair the damage done by Parkinson's disease, for example, has been greatly hampered in the United States and elsewhere by objections from anti-abortion groups that fear women will get themselves pregnant and abort their babies so they can sell the fetuses to research laboratories.

In June 1997 Harold Varmus, director of the National Institutes of Health, was grilled for three hours by a Congressional committee because an NIH-supported scientist was accused of using fetal tissue in his research on genetic mutations that cause disease. Varmus was forcibly reminded by the Congressmen that the rules against using fetal tissue are to be enforced, no matter what good the research might accomplish.

Basic research, the kind of research that seeks new knowledge for its own sake, is undoubtedly the best bargain any taxpayer has ever had. Victor Weisskopf of MIT, a science adviser to several presidents, pointed out, "The total cost of all basic research from Archimedes to the present is less than the value of ten days of the world's present industrial production."

Large projects with clearly defined goals, such as the Human Genome Project, can also produce enormous knowledge for relatively little cost. Even at $1.00 per base, the Human Genome Project's "guesstimated" $3 billion cost is far less than our annual expenditure on pizza or cosmetics. And as we have seen, the project is moving faster and costing less than the original projections predicted.

Who is paying for the research that will ultimately lead to human immortality? You are: the taxpayer. The President of the United States proposes a national budget each year, which includes funding for research. The Congress debates and decides on how much of the President's requested budget will actually be funded. There is always a tension between the White House and Capitol Hill, a tension that is usually fed by partisan politics.

Congress must vote twice on each appropriation in the budget: once to fix the amount of the funding and the second time to actually authorize spending the money.

Funding for basic research is so small in comparison to most budget items that it rarely causes a stir in these political debates unless it is for a large, visible project such as NASA's International Space Station or the Superconducting Super Collider that particle physicists wanted and failed to get.

To date, funding for biomedical research has been almost entirely free of political rancor, except for issues touching the volatile abortion question. The Human Genome Project was accepted and funded with relatively little struggle, but it helped to have someone as prestigious as James Watson willing to take the responsibility of directing the program at its outset.

Who Pays for the Elderly?

The first effects of the research that will lead to human immortality are already being felt today: People are living longer. As bioscientists learn more about the basic mechanisms of aging, as gene therapy begins to spare us from the ravages of cancer, Parkinson's, Alzheimer's, and other diseases that affect the elderly, as better nutrition and new surgical techniques help us to live longer, the number of elderly persons in our population will soar.

Who will pay for their upkeep? This question assumes that most of the men and women over age 65 will no longer be earning a living, except from pensions or insurance of one sort or another. Eventually, this assumption will no longer be valid. Even today, private savings in the

form of mutual funds, pension plans, and other invest-
ments (including housing) already rival Social Security as
a source of retirement income. Still, the question of who
will pay for the elderly must be answered.

The picture is far from one-sided, however. Data from
the *National Long-Term Care Survey* have shown that the
proportion of elderly who are chronically disabled or institutionalized in old-age care centers decreased from 23.7
percent in 1982 to 22.6 percent in 1989. This trend will
probably accelerate as research in aging and its associated
diseases helps the elderly not merely to live longer but to
live healthier, more active lives.

The Alliance for Aging Research has estimated that if
people over 65 could remain healthy and active the nation
would save some $5 billion *per month* in government and
private spending for health care for the elderly.

Today, despite private savings, taxpayers pay for most
of the care of the elderly. In 1992, 93 percent of those over
65 received Social Security benefits that averaged $6,634
per year. Hardly a princely sum. But for 63 percent of
those elderly, their Social Security checks provided at least
half their total income. For 26 percent, Social Security was
90 percent or more of their total income. For 14 percent,
it was their *only* source of income.

Some economists and politicians have suggested that
the Social Security Administration should be allowed to
invest its holdings in securities: blue chip stocks or long-
term bonds that will pay interest and dividends, thereby
increasing the amount of money available to fund the system. Opponents fear that the system could just as easily
lose money that way as make it.

While this debate rumbles on in Washington, the number of retired persons demanding their Social Security
benefits continues to rise. Between now and the year 2005,
the growth rate will be modest, because birth rates were
low during the Great Depression of the 1930s. But when

the Baby Boomers of the postwar generation that started in 1946 begin to reach retirement age, in 2011, tens of millions of additional retirees will flood the system. The current method of funding the system will clearly be inadequate.

Medicare, the government-backed medical insurance program for the elderly, is in much the same situation. The Medicare trust fund is already in danger of going broke unless costs are controlled better, benefits cut, or tax contributions increased—or some combination of all three.

Politicians find it very distasteful—and dangerous—to cut benefits or increase taxes. And in any large program such as Medicare there is bound to be waste, duplication, and outright thievery. The Inspector General of the Department of Health and Human Services reported that in 1996 the Medicare program was bilked of $17–28.6 billion through waste, fraud, or abuse.

The End of Social Security and Medicare?

Now add the new factor: By 2011, it will be quite evident that unprecedented numbers of men and women are living past their nineties, past 100, and even starting to crack the long-standing life span limit of 120.

Who's going to pay for them?

Perhaps nobody. Increased human longevity, stretching toward immortality, may very well be the straw that breaks the back of Social Security, Medicare, and private pension systems.

Today politicians are already suggesting means tests for Social Security recipients. Why pay out benefits to people who don't need them? Does a widow with a personal income of $100,000 per year really need another $6,634— even if she is entitled to it by today's laws?

Slowly, it will become apparent that the "elderly" are

still young enough, physically, to keep on working. Why retire at 65 if you are still as vigorous as you were at age 45? There will be a painful transition period and a massive amount of political squabbling, but sooner or later today's Social Security and Medicare systems will be changed beyond recognition—or swept away altogether.

Can we trust politicians or their bureaucrats to wisely and honestly administer a system that checks each individual's financial status before determining whether that person should receive pension funds derived from tax money? This may become the biggest political imbroglio of the twenty-first century's first decade, because this is exactly where current government-backed pension and medical insurance systems are heading, not only in the United States but throughout the industrialized world.

18

Can We Afford It?

I'd give all wealth that years have piled,
The slow result of life's decay,
To be once more a little child
For one bright summer-day.

—LEWIS CARROLL
(Charles Lutwidge Dodgson)

∞ THE AUTHOR OF *ALICE'S ADVENTURES IN WONDERLAND* summed up what most of us feel in his poem "Solitude": We would pay anything we have to avoid aging and inevitable death.

Despite the social problems it raises, despite the moral and religious issues, we want to cling to life. In the previous chapter, we saw that as biomedical research learns how to extend the human life span, our political and social systems for taking care of the elderly will be forced to change drastically.

What about our nongovernmental systems of business and commerce? How will they be affected by steadily increasing life spans and life expectancies?

The Future of Private Insurance

Private insurance companies will be able to adjust to these new conditions and even take advantage of them, if they are allowed to do so by the government agencies that regulate them. Private medical insurance policies, for example, could continue pretty much as they are now: The insured party keeps on paying premiums and the insurance company covers whatever medical costs are incurred (or a part of those costs).

Indeed, as research leads to control or elimination of many of the diseases and infirmities of age, health insurance companies will become more profitable because they will be paying out less for claims.

Private life insurance policies are considerably different. A life insurance policy is essentially a bet between you and the insurance company. Backed by actuarial statistics on life expectancy, the insurance company bets that you will not die before you are 65. If you do, the company pays out more than you have paid in with your premiums. After you reach 65, the bet switches: The company will pay you either a lump sum or an annual amount, betting that you won't live long enough to cost them more than the money they have collected from you in premiums.

The system works well enough, apparently. Tens of millions of Americans carry one form of life insurance or another. Insurance companies are generally profitable and have an enormous cash flow. As life spans and life expectancy increase, the insurance industry can adjust the "payoff" age from 65 upward—if the government regulators will allow them to.

There will, of course, be political pressures from the voters to maintain the age of payoff at 65; that way, the insured can stop paying premiums and the insurance company will have to pay its customers. But all that this will accomplish, in the long run, will be to bankrupt the insurance companies and leave everyone without any private life insurance at all.

Again, the situation comes down to decisions made

by politicians and bureaucrats; in this case, those who regulate the insurance industry. The short-term political pressure to "hold the line" will be enormous. People who can confidently expect to live well past 100 will see quickly enough that by keeping the payoff age at 65, they will get much more out of their private insurance policies than they have paid in.

214

But you can't take out of *any* system more than you put into it. That is a law of nature. Physicists in the nineteenth century discovered the laws of thermodynamics which, in ordinary nonphysicists' language, state that you can't win, you can't even break even, and you can't get out of the game. Whatever you get you must pay for, and you must even pay more than the value of what you're receiving. It may not be fair, but it's the law. Stars, galaxies, atoms, molecules, bacteria, Sequoia trees, you, and I are all bound by it.

TANSTAAFL, in the phrasing of science fiction writer Robert A. Heinlein: "There *ain't no such thing as a free lunch.*"

The private insurance industry will have to change or go out of business. As people live longer and longer (and healthier and healthier) private insurance systems will be forced to adjust to the new actuarial statistics.

Will immortals buy insurance? Yes. Even when the last secrets of the aging process have been uncovered and dealt with, even when human life spans extend indefinitely, there will always be the chance of accident or malevolence.

Insurance may be used for different purposes. Immortals can plan for the long term. If you expect to live for centuries, perhaps you will want to take out an insurance policy that is essentially a savings program. You pay your premiums for 100 years, and at the end of that time, you have accrued enough money to knock off work for a few decades and loaf. Or start a new career. Or study architecture. Or take a hike around the world. When the benefits run out, you can go back to work.

How Much Will It Cost?

So far we have ignored a basic factor: The medical treatments that will extend our life spans are going to cost something. How much, no one knows yet. But we can assume they will be quite expensive, at least at the outset. New medical procedures usually are.

One of the arguments that will be used against life extension will be that the treatments are so costly they can only be afforded by the wealthy. This is the populist position: "If it can't be made available for everyone, then no one should be allowed to have it."

It might be a powerful argument against early clinical trials of life-extension procedures. The populist argument usually wins a good deal of political support.

Yet this argument will lose in the long run. First, it will lose because the wealthy *will* indeed seek life-extension treatments, if not in the United States, then overseas, where the politics are more congenial and the laws more lax. Second, it will lose because everyone, rich or poor, will want the benefits of life extension. No sane and healthy person wants to die.

The pressure to move ahead with life-extension trials will overpower all resistance. Politicians will be forced to find ways to fund the work or be voted out of office and replaced by others who will do so. Massive government funding will be demanded, with the demand increasing with each apparent success.

Of course, there will also be plenty of frauds and cheats who will be perfectly happy to milk customers and/or government agencies for every penny they can get. Not only the poor are always with us; the crooks are too.

To some degree, we have a valid precedent for this situation: organ transplants.

The first successful heart transplant, in 1967, stunned the world and quickly roused demand for organ transplants of every type. The procedures were fearfully expensive at first and are still far beyond the means of the average middle-class American. Yet financial arrangements are made routinely for such procedures. When all else fails, families appeal to their neighbors and the public at large for contributions to help them.

There is no telling, at this stage of research, how much life-extension procedures will cost. In all likelihood they will not be as expensive as organ transplant surgery. On the other hand, *everybody* will want the treatment. Even if the cost is very low, the demand will be very high.

Perhaps the opportunity for immortality will encourage people to save or invest wisely so that they will have the money for the treatments when they reach their forties. On the other hand, if it is determined that the treatments work better when administered at younger ages, an entire new branch of banking may open up: long-term loans for young men and women to get their life-extension treatments. After all, they should have many, many decades in which to pay the loans back. That's the way people buy houses; why not buy immortality with a loan from your friendly neighborhood bank?

Religious Objections

Of all the barriers to life extension and immortality, religious injunctions will undoubtedly be the most severe.

Most religions teach that there is life after physical death. Moreover, in many religions the afterlife is pictured as an eternal reward (or punishment) for one's behavior in this life. Good people go to heaven; bad people are damned to hell.

If we can extend human life spans considerably beyond the Biblical three or four score years, there will undoubtedly be objections by religiously motivated people that we are trying to circumvent God's plan for us. If and when it becomes clear that we can extend our life spans indefinitely, that human immortality can be achieved here and now, these objections will rise to a howling roar.

217

By evading death, we also evade the reward or punishment due to us in the next life. Good people will not go to heaven; they will stay here on Earth. Bad people will not be punished in hell; they can go right on in their evil ways as long as they like—or the law permits.

The moral codes that religions teach are usually backed by the idea of eternal retribution in the next life. Be good or risk eternal damnation. Cease your evil ways or you will not reach heaven. If a person can live forever, why bother listening to these moral injunctions?

Of course, there will still be secular laws. And the moral codes that religions espouse certainly do help people to live in peace and harmony with their families and neighbors. Yet the power of that eternal reward or punishment, that inescapable accounting that you must face at the end of this life—that will fade and disappear when people can live forever.

Will this "thwarting of God's plan" stop people from seeking immortality here on Earth? Probably some, perhaps many, people will refuse life-extension therapies because of their religious beliefs.

But most, I think, will alter whatever beliefs they have to suit the new opportunity. "After all, perhaps immortality here on Earth is part of God's plan for us," they will tell themselves.

To paraphrase Arthur C. Clarke, the prospect of human immortality is ticking like a bomb at the cornerstone of many faiths.

Arthur Caplan, director of the Center for Bioethics at the University of Pennsylvania and founding president of the American Association of Bioethics, writes: "If being human means using intelligence to improve the quality of life, there is little basis for ethical ambivalence or doubt about eliminating genetic scourges, much as we have done with infectious diseases such as polio and smallpox. . . . Changing our biological blueprints in pursuit of longer, healthier lives should ideally pose no more of a threat than does taking impurities out of the water supply to protect human health."

Caplan was speaking about research aimed at eliminating genetic diseases. Yet much the same argument can—and will—be made about life extension and immortality.

The Social Costs of Immortality

Now let us get back to the problems of today and look at the other side of the coin: If persons in their seventies and eighties are still working, rather than retired, what does this do to the prospects of teenagers first entering the job market? Will there be enough jobs to go around?

Today employers are finding it difficult to hire new high school graduates for jobs that require even a modest level of reading or arithmetic skills. In some industries, such as insurance, companies have made a concerted effort to get retirees to return to work, because they cannot find enough entry-level youngsters with the skills to do the required tasks.

Many firms have found it necessary to pay for educating their new hires to bring them up to the reading and math skill level where they can do the entry-level jobs for which they have been hired. This costs American industry hundreds of millions of dollars each year.

If men and women no longer retire at 65 but stay at their jobs because they are strong enough and vigorous enough to continue working, where will the jobs for teenagers come from?

Those teenagers, undereducated despite their diplomas, could live forever. Our school systems are going to have to change—and change radically—to meet the new society that human immortality will create.

Will an aging—but still vigorous—population create demands for new jobs? In states where large numbers of retirees live today, the entertainment, travel, and sports industries are booming. Florida and Arizona, for example, boast more golf courses than hospitals. An aging population does not automatically mean an economic slowdown. Instead, it means a different mix of business opportunities. And jobs.

Genetic Screening

Another problem raised by our increasing understanding of genetics stems from the possibilities of genetic screening.

Today it is possible to test a pregnant woman to see if her fetus is afflicted with Down syndrome. Many couples refuse to take the test, for fear that they might discover the woman is carrying a Down syndrome baby and the wrenching emotional turmoil that such knowledge will bring. The couple would face a tragic choice: Abort the baby or live with a retarded Down syndrome child until its premature death.

This emotional turmoil will be multiplied as the Human Genome Project and other research efforts now underway make it possible to test the entire genetic makeup of men and women even before they decide to have children.

A panel appointed by the National Institutes of Health to consider screening for the cystic fibrosis gene recommended in 1997 that patients should be offered counseling along with the screening test so that they can be helped to understand what the test results mean and what their options are if they are found to be bearing the *CF* gene. Bioethicist Norman Fost of the University of Wisconsin–Madison estimates that there are presently fewer than two thousand health-care providers in the United States qualified to offer genetic counseling to the 4 million women who become pregnant each year.

Is genetic screening a lifesaving medical miracle or an invasion of privacy? Do you want to know if you have a genetic flaw that will lead to cancer or heart disease? Does your employer want to know? Or your insurance company?

Genetic screening can help individuals to learn what threats lurk in their DNA and how to avert them. If you have a recessive gene for sickle-cell anemia, for example, you are not affected by the disease. But if you marry a person who also carries that recessive gene, there is a 25% chance that each of your children *will* develop sickle-cell anemia.

At our current stage of understanding, it is possible to detect genetic problems but not yet possible to fix them. Therefore people are very sensitive about learning that they may have such problems. Many would rather not know, even though such knowledge might help them to plan their lives and lifestyles in ways that can minimize the danger.

Many fear that such knowledge, in the hands of the government or a prospective employer or an insurance company, can result in denied medical benefits, withdrawn job offers, or canceled insurance policies.

There is also the fear of a "Brave New World" scenario. In 1932, British author Aldous Huxley's novel *Brave New World* presented a picture of a future world in which parents were genetically tested by the government and children's intelligence and physical characteristics were genetically tailored to suit the government's requirements for different types of workers. Children were conceived artificially and gestated in vast biochemical assembly plants; sex was entirely recreational, rather than procreational.

221

The result was a sort of genetically programmed dictatorship, a horrifying example of what could result from studies of genetics and cloning.

In the real world, gene therapy will be necessary to save our gene pool from the growing number of individuals who carry damaged or defective genes. By keeping diabetics alive long enough to raise families, for example, we are maintaining—and even increasing—the number of diabetics in the gene pool. Unless we learn to use gene therapy to eliminate defective genes, eventually everyone will be carrying them.

Any suggestion of "improving" the gene pool, of course, raises more fears of "master race" eugenics and ethnic-cleansing pogroms that employ geneticists instead of soldiers. Who decides which genes are "defective?" Public vigilance and public control of government policy *must* be maintained if gene therapy is to be used for our benefit and not twisted into a neo-Nazi nightmare.

Despite the fears of a "Brave New World," however, as knowledge progresses, gene therapy and other procedures will eventually be able to alter or repair genetic problems. One day repairing faulty DNA may become no more difficult than bypass surgery—and much less risky.

Eventually, such DNA repair procedures will be as commonplace as setting broken bones.

But only if today's research is allowed to continue. And only if the public understands where this research can lead—and agrees to allow it to bring us there.

Public understanding is urgently needed if we are to avoid the all-embracing genetic dictatorship of *Brave New World*. It is the truth that makes us free—and keeps us free. "Knowledge is power," as Francis Bacon wrote in 1597. Only when knowledge is spread as widely as possible will the public control such power.

Funding and Politics

While the actual amount of money being spent on research that can extend the human life span and lead to immortality is so small that it has escaped the politicians' notice so far, we cannot expect this era of benign neglect to continue indefinitely.

As research results lead to clinical trials, the media and the public will gradually become aware that new techniques for extending the human life span are becoming available. Then we will see a huge public clamor and massive controversy. Public opinion, fueled by pronouncements from political and religious leaders and fanned by intensive media attention, will be split into two diametrically opposed camps.

On the one hand will be those who oppose the idea of extending human life span, just as their predecessors have opposed cloning, birth control, and even anesthetics.

Their cry will be: "It's never been done before, therefore it must be wrong, sinful, dangerous."

On the other hand will be those who want to live forever—or at least as long as they reasonably can.

Their cry will be: "I don't care whether it's right or

wrong, sinful or not, dangerous or beneficial. *I don't want to die!"*

My opinion is that those who want to extend their life spans will win, despite all opposition. The will to live, the fear of death, is enormously more powerful than the doubts and fears of the unknown territory into which this new knowledge will plunge us.

223

19

"I Always Thought It Was a Good Idea"

This generation of Americans has a rendezvous with destiny.
—FRANKLIN D. ROOSEVELT

∞ IN HIS FIRST INAUGURAL ADDRESS ROOSEVELT SAID, "THE only thing we have to fear is fear itself."

The biggest hurdle that the concept of human immortality will have to face is the fact that it is going to happen much sooner than most people think. From the White House to the Vatican, from the ivy-covered halls of academia to the tract houses of Mr. and Mrs. Suburbia, hardly anyone realizes that within their generation—within their lifetimes— the human life span is going to stretch well beyond the current limit of 120, and average life expectancies will begin to be measured in centuries rather than decades.

The pace of research is moving much faster than society's ability to absorb such radical change.

Global Consequences

There are some 6 billion human beings on Earth. In fifty years, at current growth rates, there will be more than 10 billion (see Population Growth, opposite).

POPULATION GROWTH

Nearly 6 billion human beings live on Earth today. Most population projections see that number steadily rising through the next century, creating the threat of a population explosion that could bring a global ecological collapse.

In 1650 world population was roughly half a billion. Two centuries later, 1850, it had more than doubled, rising to 1.3 billion. It took only one century, to 1950, for it to double again, to 2.5 billion. In less than fifty years, it has doubled once more, to 5.8 billion. In the next fifty years, it will swell to 10.6 billion, according to most estimates.

Global population may stabilize around that figure or even decrease, according to a study by the International Institute for Applied Systems Analysis in Austria. Population growth rates are already slowing through much of the world, including the poorer, underdeveloped nations.

However, no study of global population growth to date has factored in the possibility of greatly reduced death rates caused by radical extensions of human life spans.

Pundits have warned for nearly half a century that we face a population explosion that can despoil our world of all its resources. They paint alarming pictures of a near future in which farmlands have turned to deserts, the air is choked with pollution, fresh drinking water is as expensive as petroleum, forests have been stripped to bare earth, and the seas have been emptied of their fish.

They predict a great "die-off," where the planet's overpopulating billions simply cannot find enough to eat or fresh water to sustain themselves. Starvation, disease, and war—nuclear war—will kill most, perhaps all, the human beings on Earth.

To escape this dread future, men and women of good intentions have urged the world's people to slow their population growth. In the rich, industrialized nations, this has actually happened. Population growth in Europe, Japan, and North America has stabilized close to zero. Even in the poor, underdeveloped nations of Asia, Latin America, and Africa, population growth rates have slowed.

Birth rates are falling. But what happens when the death rate plummets down to nearly zero? What happens when human life spans are extended by hundreds of years?

Predictions of environmental doom are not new. In 1886 the Audubon Society was founded, in part, because thoughtful people feared human expansion would leave the world bereft of all birds, with "silent hedges, once vocal with the morning songs of birds, and . . . deserted fields where once bright plumage flashed in the sunlight." In 1926 the Federal Oil Conservation Board announced that the United States would run out of oil in seven years.

Paul Ehrlich predicted in his 1967 book *The Population Bomb* that "the battle to feed all of humanity is over. In the 1970s the world will undergo famines—hundreds of millions of people are going to starve to death in spite of any crash programs embarked on now. At this late date, nothing can prevent a substantial increase in the world death rate."

The dire results predicted by population alarmists have not happened. Not yet, at least. However, as Aesop pointed out more than two millennia ago, just because the boy hollers wolf needlessly does not mean that the wolf won't eventually arrive.

While there have indeed been famines in Africa and elsewhere, in most cases they have been the result of political bungling, not a failure of natural resources. In Somalia and central Africa, for example, civil wars and ethnic genocides have been the cause of starvation, not a collapse of the ecology. The rich and powerful nations of the world have intervened in most cases with food aid and even military "peacekeeping" forces to stop the slaughters and feed the hungry.

It is important to recognize that the highest birth rates around the world occur among the poorest people. Even in the wealthiest nation on Earth, the United States, the highest birth rates are among the lowest-income groups. Since life-extension treatments will at first be affordable, most likely, only to the highest-income groups, they will not immediately affect population growth. There will be some time—a decade or two, perhaps—for societies to adjust as best as they can to the new factors stemming from lengthened life span.

This will not be enough time to make a peaceful, rational, orderly adjustment to the new facts of life. There will be protests and demonstrations, political debates of incandescent heat, thunderings from pulpits and demands that life-extension therapies be banned outright and all such research prohibited. There will also be countering demands that such therapies be offered free to the entire public, that they be paid for by the government and not reserved for the rich alone. Politicians will scramble for ways to raise taxes that can be used to pay for "free" life-extension treatments.

Still, no matter how we might wish it otherwise, life extension will inevitably lead to population growth by enormously reducing the death rate.

Death Rate vs. Birth Rate

The swift population growth of the past century has largely been the result of lowering the death rate among

227

the very young. Now we are on the verge of a similar lowering of the death rate among the old.

There are some differences of kind and degree between the two situations.

For the most part, the first men and women who receive life-extending therapies will be financially self-sufficient. It seems likely that, at first, only the wealthy will be able to afford such treatments. But as life extension is made available to larger groups of people, it will still be offered to adults, not babies or children. Presumably these adults will be able to take care of themselves financially, although a percentage of them might be on some form of governmental aid, such as welfare.

Despite the charity cases, most of the adults who obtain life-extension treatments will not only be able to pay for their treatments, they will be able to pay their own way through the added years of their lives. Instead of being a drain on the economy, they may very well be a gain: industrious, productive, vigorous adults who generate wealth for themselves and their societies.

Remember, the mental image of the elderly we now carry inside our heads is a picture of feeble, failing, sickly old people. If and when life extension becomes available for the masses, the "elderly" will be no more feeble and sickly than men and women in their forties or fifties today. Indeed, current work in regeneration, nanotechnology, hormone replacement therapy, and gene therapy can result in reversing the effects of aging so that a man or woman can be physically a young adult indefinitely.

The big question is: Will these men and woman have more children? If a man, for example, receives life-extension therapy at age 50 and thereby can reasonably expect to remain physiologically at that age for a century or more, will he begin to father more children? When life-extension therapies include age regression, so our 50-year-old can be returned to the physical condition of a 25-year-old, will he want to start a new family?

When women can reverse the effects of aging, will they want to bear more babies? How many more women like Arceli Keh will want to have babies in their sixties or beyond? Menopause occurs because a woman's supply of eggs runs out—usually by age 50 or so. Will women want to reverse the effects of menopause and restore their fertility? Life extension will raise very complex reproductive choices for women.

There is some evidence that extending the life span results in reduced fertility—at least for fruit flies.

Fruit flies are ideal subjects for geneticists, precisely because their life spans are so short—only a matter of days. Many generations can be bred and studied within a few months. The results of mutations can be seen in weeks, rather than in the decades it would take for such changes to show up in human subjects.

Investigations undertaken at the University of California–Irvine divided a set of genetically identical fruit flies into two groups, which were kept under identical conditions, except that one group was allowed to breed normally, while the other group was bred only from eggs that had been laid by flies that were near the end of their reproductive lives: elderly mothers.

This experiment was repeated for fifteen generations, although the results began to be apparent after only a few generations. The normal, or control, group showed the same life span and fertility throughout the experiment. The flies bred from elderly mothers, however, lived longer but produced fewer eggs than their normal cousins.

Each generation of the longer-lived flies was allowed to breed only from eggs laid as late in life as possible. Each succeeding generation lived longer than its parents—and produced fewer eggs.

Within a few generations, the flies of older mothers

were living about 30 percent longer than the control group, and the life span gap between the two groups was still growing when the experiment was ended.

The longer-lived flies appeared normal in size and behavior, but they could fly longer and were more resistant to starvation and lack of water than the control group.

The California experiment indicates that by delaying reproduction until late in life, the flies attained longer life spans—at the cost of fertility. They had fewer offspring.

Human beings are not fruit flies, of course. But as human life spans are extended, will there be an effect on fertility? Today's Generation X couples are already having babies at considerably later ages than their parents did. Will the lengthening human life span encourage couples to postpone starting their families until much later in life? And if so, will humans show a similar tradeoff of lower fertility for longer life span?

Will Marriage Survive?

An allied question is: What happens to the institution of marriage when people begin to live for centuries? What happens to "till death do us part" when death is put off for centuries or perhaps forever? In the twentieth century divorce rates have skyrocketed, in large part because people are living long enough to want to change partners. Death does not part these married couples; boredom or infidelity or simply a gradual estrangement over the years sunders their marriages.

This is the key to the population growth problems that life extension raises. Will men and women who live for centuries have so many additional children that they raise the global birth rate to the bursting point?

It would be logical to offer a trade-off. Life-extension therapy would be allowed only if the individual agrees not to have more children. Logical—but probably not to be expected. Such coercive programs rarely work. The People's Republic of China instituted very strict birth control measures, but they have been only partially successful. Chinese couples have either circumvented the birth control laws or paid the heavy fines imposed on those who have more than one child. The urge to procreate cannot be stifled by laws.

Moreover, in most of the world, governmental interference in family planning is regarded as anathema. Several major religions prohibit family planning and abortion altogether.

Over time, social and religious tenets will adapt to the new conditions of life, as they have adapted in the past to startling new ideas such as anesthesia and evolution and the feminist movement. But the adaptation period can be painful and fraught with dangers.

In the long term, if life extension works against human society, if it causes more disruption and pain and problems than society is willing to bear, the gift of immortality will be rejected. We will live as we have since time immemorial, facing a rendezvous with death after "threescore years and ten . . . [or] fourscore" years.

On the other hand, if life extension truly turns out to be beneficial to human society, if its merits outweigh its problems, then the gift of immortality will be accepted, and the human race will enter a fundamentally new era, free of the fears of disease and disability that come with old age.

It will take many decades to come to that decision. Perhaps centuries of turmoil and conflict are in store for us. It could be an ugly time. At best, it will be a time of strife and tension.

20

Three Futures

All this will not be finished in the first one hundred days.
Nor will it be finished in the first thousand days, nor in
the life of this Administration. . . . But let us begin.
                                                —JOHN F. KENNEDY

∞ THE SEARCH FOR IMMORTALITY BEGAN FAR BACK IN THE mists of prehistory, when early humans first realized that they would die.

During the Ice Ages, Neanderthals buried their dead with flowers and jewelry to adorn their bodies. Egyptian pharaohs and Chinese emperors had themselves entombed with all the wealth and servants they would need for another life. This quest for a life beyond death has given rise to great literature, such as *The Epic of Gilgamesh*. It has also been a key factor in the development of religion. Christianity swept the Roman Empire in large part because it specifically promised its believers life beyond death.

Today the quest for immortality is taking place in research laboratories around the world. Gene therapy, human growth hormone treatments, hormone replacement therapy, telomerase, antioxidants, tissue regeneration, nanotechnology, and a host of other investigative avenues are being pursued by dedicated men and women.

Most of them do not realize that the end product of their researches will be human immortality, any more than Rutherford and Einstein realized that one result of their studies of the atom would be Hiroshima.

Gradually, at first, the work now taking place in research labs will teach physicians how to extend human life span. The old limit of 120 years will be surpassed; not by much, at first. *The Guinness Book of Records* will be rewritten once, twice, and then just about every year. New breakthroughs will stop the ravaging effects of aging and then begin to reverse them. People will be physically young as long as they live, and they will live for centuries.

233

What biomedical discoveries will be made in the next fifty years? The next hundred? The next few centuries? You may very well live to see them—and benefit from them.

If today's research is allowed to continue. And *if* the results of such research are made available to all. Those are gigantic *ifs*.

We face three possible futures, as far as life extension and immortality are concerned. In one, the researches now under way are suppressed. In the second, they are controlled by an elite group. In the third, they are shared with the entire world.

"They Are Playing God"

In the third century B.C. the Chinese philosopher Chuang-tzu warned: "Banish wisdom, discard knowledge . . . for much knowledge is a curse."

It is entirely possible that governments around the industrialized world, which are the main funding sources for modern scientific research, will agree to stop funding studies of life extension. Witness the furor over cloning and the immediate response of religious leaders to new breakthroughs in almost any area of research that has to do with reproduction.

Many people fear new knowledge. They succumb to the catchphrase that scientists are attempting to "play God." Actually, the fear of new knowledge is often based in the fear that those who possess the new knowledge will gain unwarranted power over those who do not possess it.

234

The easiest thing to do when faced with a new situation is to ignore it. If you can't ignore it, then perhaps you can get rid of it. As the American writer Wallace Kaufman put it, "Except when in real danger, most people prefer what they know to what they must learn."

It is perfectly true that extending the human life span toward immortality will bring enormous social, political, economic, and even moral changes. The easiest way to deal with these changes is to prevent them from happening. The easiest and most certain way to prevent them is to stop all research in these areas and ban all future studies.

In the early 1970s, when molecular biologists first developed the techniques for gene-splicing, they realized that they were on the edge of a vast new era. They saw, long before anyone else, that by developing techniques for removing genes from cells and replacing them with other genes, even genes from other organisms, they were stepping into potentially dangerous territory. Suppose they accidentally produced a mutated bacterium that got loose from their laboratories and started a man-made plague?

So the world's top biologists called for a temporary halt in such research until they could devise safeguards for their research: both physical safeguards to prevent the accidental escape of genetically altered organisms and moral safeguards to guide and direct the course of future research.

Although many politicians wanted to impose additional government restrictions on the molecular biologists, for the most part governments agreed to the biologists' self-imposed safeguards and the research was resumed. But not without intense government oversight.

As today's research on life extension begins to produce clinical results, pressures to control such work will arise. Extremists will excoriate scientists who are "attempting to play God." Politicians will listen to those extremists in direct proportion to the number of votes the extremists can deliver on Election Day.

235

Suppose the extremists win? Suppose the government shuts down all the research programs that are now leading to extending our life spans and prohibits all such research in the future?

If the shutdown is imposed only in the United States, scientists will migrate to those nations where the research is permitted. This is as natural as a housepainter looking for houses that need painting. You go where the jobs are. Or you find a different line of work. Asking a molecular biologist or an embryologist to find a different line of work is like asking Tiger Woods to give up golf or asking Stephen King to quit writing.

The United States will suffer a brain drain. Our best biologists will find positions overseas. When their research pays off, its results might not be legally available to American citizens. Even if they are, we will have to pay other nations or foreign corporations for the benefits of life-extension therapies. The U.S. economy will suffer.

Suppose, instead, that *all* the nations of the world ban research on life extension and no one anywhere on the planet is allowed to pursue studies that might lead to extending the human life span. This could lead to one of two different results, depending on how strictly the ban is applied.

If the restrictions are fairly loose, scientists will continue their work—one way or the other. Studies of telomerase, for example, will not mention that such work might lead to understanding the cellular clock that controls aging. Research in regenerating tissue will not state that it might be possible to regrow damaged organs. Scientists are not fools (despite their depictions in many motion pictures and television shows). They know how to "write around" restrictions imposed upon them.

But let us assume that government bureaucrats are not fools either. They will interpret the restrictions on research quite strictly. *Anything* that might lead to increasing human life span will be barred.

If all research that might lead to extending human life span is stopped, then virtually all the biomedical research under way today will be banned. If the restrictions on research are interpreted very strictly, hardly any research program now in existence will be safe.

Then you can look forward to a future in which you may very well come down with Parkinson's disease or Alzheimer's. Cancer will lurk within your cells, waiting for the right combination of genetic damages to strike you down. You will age and die after some three or fourscore years—at most.

In truth, many people will prefer this future. It is almost exactly like the past, and most people prefer the devil they know over the devil they have yet to meet. After all, society has been going along for thousands of years with the Biblical three or four score years as our best hope. Why change? Why reach for forbidden fruit?

"Knowledge Is Power"

Think of the power that could come from immortality or "merely" life spans measured in centuries instead of decades. If Genghis Khan had not died, all of Europe and

Asia might well have become part of his Mongol empire, and the rebirth of knowledge and learning that we call the Renaissance might never have come to pass. What would the world be like today if Joseph Stalin were still alive and vigorously pursuing his megalomaniacal dreams with all the ruthlessness and terror he wielded over Soviet Russia?

Think of the possible benefits as well. What would the world be like if the saintly Albert Schweitzer were still alive and influencing world opinion? Or if Albert Einstein remained at the height of his intellectual powers even to this day?

The Morlocks and the Eloi

Suppose some government leaders decided to keep some research on life extension going, in secret, so that they could prolong their own life spans and perhaps the life spans of those near and dear to them?

We run into a science fiction scenario in which the world's government or corporate leaders and their chosen elite are immortal (or nearly so), while the vast majority of the world's population live and die from one generation to the next. That might be the most likely outcome of a world in which research on life extension is officially banned but secretly continued.

The world will divide into the immortals and the death-bound. Very likely the immortals will disguise their great ages; they will not want the death-bound to realize that they are immortal, for fear of a revolt. It should be relatively easy to do, if the immortals have control of the world's institutions of learning and the media.

In his 1895 classic *The Time Machine*, H. G. Wells portrayed a far-future world in which human society had divided into the beautiful Eloi, who spent their days lazily playing in sunny gardens, and the grotesque Morlocks, who lived underground with the machines that made everything the Eloi needed. The Eloi seemed to be in paradise, except that at night the Morlocks came to the surface and dragged a few Eloi down below to eat them.

Wells was warning against the divisions between rich and poor that he saw in late Victorian society. In a future in which immortality is available only to a few elite individuals, the world will truly divide into the Eloi and the Morlocks, and the Eloi will live in constant dread of being discovered by their death-bound "inferiors."

Or would they dread discovery, after all? Ancient Egyptian society endured almost unchanged for more than two thousand years, with the common people worshipping a Pharaoh who was regarded as a god. The Chinese empire lasted almost as long, with interruptions from barbarian invaders from time to time, also with a god-emperor at its head.

Could it be that the immortal elite will not hide their longevity but revel in it? To twentieth-century Americans, accustomed to the benefits and outlook of democracy and equality of rights, it seems unlikely. But most of the world has not had our democratic experience.

The great nineteenth-century Russian novelist Fyodor Dostoyevsky observed in *The Brothers Karamazov* that mankind wants "miracle, mystery, and authority" rather than freedom. "So long as man remains free he strives for nothing so incessantly and so painfully as to find someone to worship."

Who better to worship than god-kings who are truly immortal? To those who have never heard of Jefferson or Lincoln, an eternal empire led by demonstrably superior immortals might be much more reassuring than the cantankerous, clamorous, constant squabbling that is democracy.

Immortality for All

If there is one lesson that the twentieth century has taught, however, it is that people seek freedom. Despite the horrendous dictatorships of Hitler and Stalin and lesser tyrants (or perhaps because of them), the wave of the future appears to be democracy based on individual freedom.

As life-extension research begins to pay off and clinical results show that men and women can extend their life spans for centuries, with no absolute limits in sight, the clamor for life-extension treatments will be undeniable.

239

This is why it is so vitally important to keep this research in the open, to report its results fully and widely. As long as the general public knows that such work is under way and understands its implications, it will be impossible for governments or corporations to make immortality the secret preserve of an elite few—unless the general public agrees to officially prohibit such research. An official ban will inevitably lead to secret continuation of the research for the benefit of an elite few. The temptations are too tremendous to assume otherwise.

In all likelihood, though, despite religious and moral objections, despite economic dislocations and political upheavals, virtually everyone on Earth who learns about the possibilities of immortality will want its benefits. Society will be changed by this, perhaps beyond recognition. Governments will be pressed to find ways to make life-extension treatments available to all their citizens. Fortunes will be made by those who can provide the materials and skills for the various therapies—unless governments step in and control prices, supplies, and manpower.

The wealthy, industrialized nations of the world will undoubtedly be the first to have large numbers of their citizens extend their life spans. This can lead to a new global division, in which the citizens of the richest nations begin to extend their life spans a generation or more before such treatments are widely available to the citizens of the poorer, less developed nations.

For a while, perhaps several generations, the major division in world politics will be between the rich nations, filled with long-lived elders, and the poor nations, teeming with young people who have not yet received life-extension therapies.

This will exacerbate a trend already present in today's world: The industrialized nations have largely controlled their population growth but are growing older and grayer, while the developing nations are bursting with teenagers who are busily begetting more children. The most potent and dangerous drug in the world is testosterone: It is a major driving force behind today's terrorism and crime.

A friend of mine who has spent much of his life in the Middle East told me that the typical boast of a teenaged male was that by age 16 he had "killed his man and begotten his man."

Yet in the long run, a world in which life extension is available to all will come to pass. It may not be an unmixed blessing, especially at first.

Essentially, life extension lowers the death rate close to zero. If we are not willing or able to lower the global birth rate, life extension will lead to steady population growth, perhaps the kind of population explosion that could cause the environmental collapse and global strife envisioned in the harrowing 1972 study, *The Limits to Growth*.

Let us assume that we learn to solve the population growth problem. After all, there is some research evidence that longer-lived animals tend to have fewer offspring than shorter-lived ones. And we do have human intelligence working for us. We are not poor dumb brutes like the dinosaurs, doomed to extinction by a roll of the cosmic dice that slammed a meteoroid into the Earth. We can think, plan, foresee.[23]

One of the great benefits of life extension will be that it will allow us to grow wiser with experience. Age is the great teacher, and human beings who can separate long life from the physical debilitations of aging should be able to learn and grow wiser with their accumulating years.

Writing in 1954, the anthropologist Carleton S. Coon summarized humankind's prospects for the future in these words:

> *A half-million years of experience in outwitting beasts on mountains and plains, in heat and cold, in light and darkness, gave our ancestors the equipment that we still desperately need if we are to . . . live happily ever after in the deer-filled glades of a world in which everyone is young and beautiful forever.*

That future world awaits us if we are, as Coon believed, smart enough to earn it.

Long-Range Responsibilities

One great advantage of a world in which human life spans are measured in centuries is that it will finally allow us (or force us) to tackle the truly long-term problems that we face. Today most people hardly ever think about the long-term future. "Why worry about global warming if it's not going to have any real impact until after I'm

---

[23]Even today we are taking the first steps toward spotting potentially threatening meteoroids and laying plans for methods of diverting them away from Earth.

dead and gone? Who cares about budget deficits? Or population growth?"

The long-term problems of our environment, of our racial relations, of government deficits, of the economic disparities between rich and poor—people who live for many centuries will not be able to ignore them or pass them on to future generations.

Immortality will bring not only wisdom but responsibility. The human race will end its adolescence and attain true adulthood at last.

In Shakespeare's words:

How beauteous mankind is! O brave new world,
That hath such people in't!
—The Tempest, V, i, 183

242

21

Christmas Yet to Come

"Before I draw nearer to that stone to which you point,"
said Scrooge, "answer me one question. Are these the
shadows of the things that Will be, or are they the shadows
of the things that May be only?"

—CHARLES DICKENS,
A Christmas Carol

∞ I FEEL LIKE THE GHOST OF CHRISTMAS YET TO COME. I
have tried to show in this book the possibilities of the
future, what might happen if current research is allowed
to reach its fruition. Yet I know full well that there is no
single, specific, preordained future. The future is as open
and wide as the sea. It is created, moment to moment, by
what we do—and what we fail to do.

We can build a future in which the human life span
is extended indefinitely. There are many who fear such a
possibility and will strive to prevent it from coming about.
Yet the human spirit will not be denied. What our minds
can conceive, our hands will eventually build. The naysayers may delay the moment when human immortality
is achieved, but they cannot block it forever.

Science as Revolution

Human institutions are inherently conservative. Law, religion, social customs, all human institutions are rooted in

the need to provide a steady, stable base for society's interactions. Like biological organisms, human societies try their best to avoid mutations and keep their basic forms intact.

Thus human institutions, such as religion and law, education and ethics, attempt to maintain society's status quo. They are devised (or inspired) for the purpose of keeping tomorrow just the way yesterday was. That is why they exist. They are necessary, and every society in human history has had its protective institutions.

There is one human institution, though, that is not conservative. That institution is science. By its very nature, scientific research is always changing society by discovering new things, inventing new ideas. While all other institutions are essentially backward-looking, attempting to preserve the past, scientific research is inherently forward-looking, searching into the future, trying to find out what might be over the next hill.

The terms *backward-looking* and *forward-looking* should not be taken pejoratively. Society needs its conservative institutions; without them, society would disintegrate.

But the differences in aims and attitude between science and all the other human institutions explain why scientific research seems so often to be at odds with society's beliefs. Scientists find new realities about the world in which we live, and these discoveries inevitably change society. All our other institutions attempt to hold fast to the norms of society and resist change.

Religious believers of many faiths have opposed scientifically driven changes in society for millennia. Today much of the resistance to new scientific ideas stems from their seeming conflict with religious credos. We have seen the example of cloning, where the first success with a sheep immediately stirred a worldwide storm of protest against human cloning.

There will be religiously based objections to extending

the human life span. That is to be expected. Yet there is no discernable prohibition in any of the world's major religions against scientific research in general or research aimed at extending the human life span in particular. The objections come not from the Torah or Bible or Koran or Vedas, but from the *interpretations* of scriptures made by individual men and women.

However, there is a crucial New Testament passage that might be interpreted as encouraging—even demanding—that we do everything in our power to enhance our knowledge and capabilities.

In the Sermon on the Mount, Christ exhorts the multitude, "Be ye therefore perfect, even as your Father which is in heaven is perfect." (Matthew 5:48)

Might we dare to interpret this as a command to become the best we can, not only morally but physically, mentally, emotionally? Might not Christ be telling us that we are expected to build heaven here on Earth, to reach for perfection here and now? Might not heaven be a place and a time that we build for ourselves, the way an eager, intelligent child builds a sand castle on the beach under the watchful eye of his protective father?

Of course, that is only one person's interpretation of one passage in the New Testament. But human perfection is what we should be striving for, and scientific research into life extension and every other facet of the natural world is one of our best tools for reaching toward such perfection.

Revolutions, Future and Past

The prospect of human immortality may seem strange, even frightening. Yet it is within sight. Some who read this book will live for centuries and more. Death will no longer be inevitable.

That future is no more than a few decades away; fifty years—at most. Based on what we know today, the next half century will see a revolution in biology and medicine, which will in turn cause revolutions in society, politics, economics, ethics, and religion.

There have been several such revolutions earlier in human history.

The discovery of fire transformed humankind from just another species of clever ape into a world traveller who not only survived the Ice Ages but flourished in them. Fire gave us a source of energy beyond our own muscles or the muscles of domesticated animals.

The invention of agriculture produced a population explosion and civilization. The old ways of hunting and gathering largely disappeared as humans produced food out of the ground, built cities, and invented writing. Tribes gave way to kingdoms and empires. And tyrannies.

Ten thousand years later, the first industrial revolution, based on steam power, led to a further population explosion and put an end to human slavery. Steam-driven machines were cheaper and more productive than slaves; they provided the economic basis for what the moralists had been preaching for centuries and led to "a new birth of freedom," to use Lincoln's words.

Within our own lifetimes, a second industrial revolution—this time based on electronic chips—is transforming us into an information society and making the world into a global village.

Looking Backward

Now we face the prospect of human immortality. At first it may seem to be a long way off, but the chances are that it is no more than fifty years away. If that seems too optimistic, look back at what biomedical research has accomplished over the past half-century.

In 1950 most American homes did not yet possess a television set, the Brooklyn Dodgers were still in Brooklyn, and the Korean War broke out. The first medication for high blood pressure, a tranquilizer called reserpine, became available.

No one knew that mitochondria exist; those cellular power plants were discovered in 1951. Cancer patient Henrietta Lacks donated cells from her cervical tumor for research in 1951; her HeLa cells have been reproducing ever since.

In 1950 polio was a national scourge each summer. The Salk vaccine was introduced in 1953. In that same year researchers uncovered the structure of a protein (insulin) for the first time. And a heart-lung machine was first used in surgery.

The structure of the DNA molecule was discovered by Watson and Crick in 1953, opening the way to molecular biology and genetic engineering. Efforts began to decipher the genetic code—the method by which genes produce proteins.

The Korean War ended in 1953.

In 1953 no one knew what viruses were made of or the fact that ribosomes are the organelles within cells that actually manufacture new protein. Virus structure was determined in 1955; ribosomes were discovered the following year.

In 1957 the U.S.S.R.'s *Sputnik* became the first artificial satellite to orbit Earth.

In 1959 Feynman suggested the possibility of molecular engineering; twenty-seven years later Drexler coined the term *nanotechnology*.

Messenger RNA was discovered in 1960, the same year that birth control pills went on sale in the United

States and an American U-2 spy plane was shot down over the Soviet Union.

In 1961 the Berlin Wall went up, American-backed Cuban rebels were disastrously defeated by Castro's forces at the Bay of Pigs, and bioscientists discovered that cancer stems from mutations in DNA. Also in 1961 Yuri Gagarin of the U.S.S.R. became the first human to fly in space.

248

John Kennedy was assassinated in 1963.

The first laser eye surgery took place in 1962, the first home equipment for kidney dialysis became available in 1964, and the complete structure of transfer RNA was discovered in 1965. Also in 1965 studies found that estrogen hormone replacement therapy significantly extended postmenopausal women's life expectancies.

The American troop buildup in Vietnam rose toward the half-million mark in 1965, the year that Medicare began. A vaccine for German measles was introduced in 1966. The first heart transplant, first coronary bypass operation, and first use of mammography to detect breast cancer all occurred in 1967.

Race riots struck more than one hundred American cities in 1967. Robert Kennedy and Martin Luther King, Jr., were assassinated in 1968.

By 1968, the year of the Viet Cong's Tet Offensive, the complete genetic code had been deciphered. The following year a single gene was isolated, and in 1970 an artificial gene was constructed.

In 1969 astronauts Neil Armstrong and Buzz Aldrin landed on the Moon.

By the year 1970, scientists still did not know that somatic cells are subject to the Hayflick limit and cannot reproduce indefinitely. Nor did they understand that the chromosomes' telomeres shorten with each reproduction. The first CAT scan was performed in 1972, the same year that the term *apoptosis* was coined for the process of programmed cellular death.

The first gene-splicing experiments, which opened the door to genetic engineering, were done in 1973. The next year, while the Watergate scandal forced President Nixon to resign, leading molecular biologists called for a voluntary halt in gene-splicing experiments until proper safeguards could be devised.

The United States quit Vietnam in 1975.

249

In 1976 Genentech became the first company formed for the purpose of developing genetically engineered products.

In 1977 smallpox was eradicated. Smallpox bacteria no longer exist in nature. The first balloon angioplasty for clearing plaque-clogged arteries was performed that same year.

In 1978 the first "test tube baby" was born, conceived by artificially inseminating an egg *ex vivo* and then implanting the fertilized egg into the mother.

By 1980, the concept of glucose browning as a contributor to the aging process was still unknown. The AIDS virus was first diagnosed in 1981. The first commercial product of genetic engineering—human insulin produced by altered bacteria—went on the market in 1982. In 1986 the U.S. Food and Drug Administration approved the first vaccine—for hepatitis B—developed through genetic engineering.

Also in 1986 the Human Genome Project was proposed. The following year James Watson agreed to be its first director.

In 1988 the first U.S. patent for a genetically altered animal was issued to Harvard University for *oncomice*, laboratory mice altered to carry human cancer genes. DuPont Corp. started selling oncomice to research institutions.

In 1989 the Berlin Wall came down.

Also in 1989 *CFTR*, the gene for cystic fibrosis, was

identified, and the structure of telomeres was first determined.

In the decade of the 1990s, the Soviet Union collapsed, environmentalists warned somberly of a global green-house warming, and research results in biology and medicine seemed to be accelerating.

The first successful gene therapy was accomplished in 1990, repairing a damaged gene that caused severe combined immunodeficiency in 4-year-old Ashanti DeSilva.[24] Also in 1990 the protein p53 was shown to be present in a mutated form in about half of all human cancers, while unmutated *p53* genes suppressed the growth of cancer cells. In 1993 the *p53* gene was shown to be associated with programmed cell death (apoptosis).

In 1994 the *obese* gene was linked to diabetes, and researchers discovered that somatic cells do not activate telomerase when they divide, while cancerous cells do.

In 1996 experimenters reported partial regeneration of severed nerve cells.

In 1997 laboratory mice were reportedly cured of cystic fibrosis in utero.

Also in 1997 a sheep was cloned from an adult ewe's cell, and researchers announced they had regenerated several organs of laboratory mice, rabbits, and sheep.

The Perfectibility of Humankind

Considering all that has been accomplished in the past half-century and the fact that scientific knowledge appears

---

[24]Ashanti DeSilva had inherited a defective gene from both parents that failed to produce *adenosine deaminase*, which is necessary for the immune system's proper functioning. Although she still needs occasional booster treatments, she is an otherwise normal, healthy 11-year-old.

to be growing exponentially, the next half-century should see human life spans extended to centuries—indefinitely, in point of fact. For all practical purposes, aging will be a thing of the past, and death will be an option rather than an inescapable end to life.

The stage has been set for a new era of human accomplishment. This new age will not come painlessly. No fundamentally new era does. But it will come.

"Be ye therefore perfect, even as your Father which is in heaven is perfect."

We are striving for perfection.

Two hundred years ago the French mathematician and philosopher the Marquis de Condorcet wrote, "Human institutions . . . [are] capable of improvement as we become enlightened . . . nature has placed no limits to our hopes." He was hardly naive; he wrote his treatise on the perfectibility of humankind while being hunted by agents of the French Revolution who wanted to execute him.

Yet he firmly believed that we can move toward perfection. History has shown that despite the wars and slaughters of past and present, his optimism is not unjustified.

The first immortals are already living among us. You might be one of them.

I like the dreams of the future better than the history of the past.

—THOMAS JEFFERSON

The only limit to our realization of tomorrow will be our doubts of today.

—FRANKLIN D. ROOSEVELT

Appendix A: Bacteria

∞ While some forms of bacteria are harmful to us, most are not. Many are even helpful. Regardless of their relationships to us, bacteria must honestly be regarded as the most successful life-form on our planet.

Microscopically small bacteria live in every nook and cranny on Earth: on land, sea, and in the air. They live wherever there is water: inside our bodies and in the bodies of all other animals, on a damp sponge, at the bottom of the ocean, even on dust motes blowing in the wind. Any organic matter that has the slightest trace of dampness in it will quickly be attacked by bacteria; that is what makes food spoil if it is not salted or refrigerated or protected in some other way. Bacteria thrive in steaming geysers and in Arctic ice and in boiling hydrothermal vents miles below the surface of the sea. Some species live deep underground in utter darkness, digesting minerals from solid rock to generate the energy they need to keep on living.

In 1996 NASA scientists examining a meteorite that originated on Mars discovered tiny mineralized structures they believe are the fossil remains of bacteria that lived underground on Mars some 3.5 billion years ago. Exobiologists now speculate that bacterial forms of life may exist underground on many worlds—such as the frozen moons of Jupiter—where conditions on the surface have long been thought too harsh to support life.

As University of Massachusetts biologist Lynn Margulis points out, bacteria "can swim like animals, photosynthesize like plants, and can cause decay like fungi. One or another of these microbial geniuses can detect light, produce alcohol . . . fix nitrogen [for fertilizing plants], ferment sugar to vinegar . . . because their survival imperative led to their inventing every major kind of metabolic transformation on the planet."

253

In fact, it was the blue-green cyanobacteria that transformed Earth's early atmosphere of choking carbon dioxide and methane into the oxygen-rich air that we breathe. If bacteria had not done that, our form of life could not have arisen.

Margulis points out that: "Any organism, if not itself a live bacterium, is then a descendant—one way or another—of a bacterium. . . ." That includes us. The 100 trillion cells of our bodies are descended from bacterial cells.

Single-celled bacteria are tough. They have to be. They must protect themselves against the rough and dangerous environment outside their bodies. They must maintain within their bodies the delicate balance of chemicals and conditions that allows them to function. Within their single cells, they must perform all the functions of a living creature: finding food, metabolizing it into energy, eliminating wastes, avoiding predators, and reproducing.

Such cells are far from simple. In fact, they are marvels of compact complexity. A bacterium can perform all the functions that you do, even the equivalent of tasting and smelling, all within its one cell.

Some bacteria propel themselves through their watery environment with a flailing *flagellum* (from the Latin word for "whip") that can rotate some twelve thousand revolutions per minute. A bacterium can swim ten body lengths in a second, the equivalent of a human running a mile in one and a half minutes. The flagellum is connected to the bacterium's body, incidentally, by a rotating hub; bacteria invented the wheel more than 3 billion years ago.

The bacterium must keep its interior *cytoplasm*, the protein-rich liquid that fills the cell, separated from the harsh and dangerous outside world. To accomplish this, the bacterium's outer membrane is, in the words of the American author Bruce Sterling, "a slimy, lumpy, armored bark . . . the outer cell wall [is] a double-sided network of interlocking polymers . . . almost like crystalline layers of macromolecular chain mail, something like a tough plastic Wiffle ball."

That tough, wrinkled hide protects the cell's interior machinery from the hazards of the outside world. But it is not impervious. Rather, it is studded with pores made of complex proteins that protrude from the cell's outer membrane like snub-nosed cannons sticking out of the hull of a pirate ship. These pores suck in molecules from the world outside the bacterium and spew out waste molecules from the cell's interior. They are powered by specialized electrochemical pumps that can ingest enough nutrients from the outside environment to double the bacterium's size in about twenty minutes when sufficient nutrients are available.

Between the bacterium's outer hide and its inner cell membrane is a mix of special proteins and digestive enzymes called the *periplasm*. Chemoreceptors in the periplasm can sense food, somewhat like our own senses of smell and taste. Digestive enzymes in the periplasm begin the process of assimilating nutrients, once they have been sucked inside through the pores. They break up the molecules of food, doing the job chemically that our teeth do mechanically.

Other pores eject the waste products of the cell's metabolism. Thus the bacterium moves through the water (or the liquid interior of your body) in a greasy cloud of its own waste products.

254

Appendix B: DNA

∞ In the molecular world, DNA truly is a giant. While complex organic molecules may contain thousands of atoms, DNA consists of tens of millions of atoms.

The backbone of the DNA molecule consists of two long spiraling strands that are composed of chains of phosphoric acid (compounds based on the element phosphorus) and sugars (carbohydrates built of carbon, hydrogen, and oxygen atoms). Whenever I hear a mother tell her child not to be afraid of getting wet in the rain because, "You're not made of sugar," I think, "Ah, but we are."

The phosphate-sugar strands form the twin backbones of DNA's double helix, the two outer tracks of the twisting ladder. The interesting part of the molecule, though, is the cross-links that hold the two intertwining backbones together, the "rungs" that connect the tracks. These links are made of nitrogen compounds known as *bases*.

There are some 3 billion bases in the forty-six chromosomes of each human cell, 3 billion rungs on the chromosomal ladders. Yet, these 3-billion-some bases come in only four varieties. The bases are called adenine, thymine, cytosine, and guanine. They are usually abbreviated into their initial letters: *A, T, C,* and *G.*

Each of the bases is anchored to one of the coiling phosphate-sugar units of the molecule's twin backbones. This combination is called by molecular biologists a *nucleotide*. Thus a DNA nucleotide is composed of one phosphoric acid, one sugar, and one of the four bases: A, C, G, or T. Essentially, a nucleotide is a unit of one of the double helix's outer tracks, plus one-half of one of its interconnecting "rungs."

On their other ends, each nucleotide is linked to another base that is anchored onto the opposite spiral backbone of the double helix. The end of one base latches onto the end of another, like two pieces of a jigsaw puzzle. They are called *base pairs* when they link together. The two spiraling phosphate-sugar "tracks" of each DNA molecule are joined to one another by some 60 million base pairs.

However, the link between the base pairs is, chemically speaking, rather weak. Moreover, these base pairs are very specific in their choices of partners. Adenine will link *only* with thymine, and cytosine links *only* with guanine. Thus, in every cell of every creature that lives on Earth, the bases pair A–T, C–G or T–A, G–C. No other base pairings are possible.

On that simple yet fundamental linkage lies the secret of life.

Replication and Manufacturing

To reproduce itself, or *replicate*, the double helix unzips itself. The base pairs split apart and the molecule unravels into two long strands with unmated bases exposed. Meanwhile, unattached nucleotides are floating in the cell's nucleus. These are linked to the unmated bases of the DNA molecule's exposed strand in exactly the same order as the bases were originally paired up: A with T, C with G.

Appendix B: DNA

No other combinations are possible. At the end of this replication we have two identical DNA molecules, each with one of its original spirals and a new one that is exactly like the one it formerly was linked to.

This is the secret of genetic inheritance. It is how cells reproduce themselves. First the blueprint (the DNA) replicates itself. Then the cell can reproduce a faithful copy of itself. If there is a mistake in the DNA replication or if the DNA is damaged—if the genes do not copy themselves exactly—the daughter cell will be different from its mother.

In addition to replicating itself, DNA also directs the manufacture of new protein. Proteins are the basic material of our bodies, as fundamental to us as concrete is to a building contractor. The cells of our bodies, the many hormones and enzymes that our glands produce, are all mostly proteins. Proteins are us.

Genes are groups of bases along the DNA molecule. They vary in length: Some genes are a few thousand bases long, others 100 thousand or more. The genes direct the cell's production of proteins; the typical human cell contains about ten thousand different proteins. Genes carry the master blueprints for all the proteins that our bodies are constantly manufacturing, every moment of our lives.

To build new protein, the double-helix molecule unzips only partially, exposing only a particular gene—a stretch of a few thousand to 100 thousand or so of its millions of base pairs. The exposed bases are met and matched by free-floating nucleotides that have been carried into the nucleus. Following the pattern of the exposed bases, the nucleotides link together with them. Then the resulting string of nucleotides pulls away from the DNA.

What has been formed is a molecule of messenger RNA (mRNA). RNA is ribonucleic acid, a molecule that is very similar to DNA, the main difference being that RNA has only a single strand of phosphate-sugar back-

257

bone instead of DNA's double helix. The RNA molecule also has four bases anchored to its backbone. Three of the bases are the same as DNA's: adenine, guanine and cytosine. Instead of thymine, however, RNA contains a similar unit called *uracil*.

In making a protein, the DNA's guanine links with the mRNA's cytosine, just as it does in DNA replication. The adenine in the DNA, however, links with the uracil of the mRNA.

The sequence of DNA bases serves as a template that builds a new messenger RNA molecule by linking up bases like the units of a Tinkertoy set. Thus, a sequence of bases in the DNA molecule that reads C–C–G–T–A–A will produce a section of mRNA molecule that reads G–G–C–A–U–U.

The mRNA molecule has transcribed the order of bases from the DNA and carries this as a blueprint for the manufacture of a specific protein. It leaves the cell's nucleus and heads for one of the ribosomes. The ribosomes are the cell's assembly shops. In the ribosomes are molecules of *transfer RNA* (tRNA), each of them toting an amino acid molecule. Amino acids are the building blocks of proteins.

There are twenty amino acids, yet they can be combined to form thousands of different proteins. Amino acids can be thought of as the words of a very simple language which can be combined in many different ways to form thousands of different sentences.

In the ribosome, tRNA carries in the "raw materials," the amino acids. The mRNA brings in the blueprint for building a new protein molecule out of the amino acids. Following the blueprint etched into the mRNA, amino acid molecules are joined together to form a new protein, linking up in the order prescribed by the mRNA, like the engine, coaches, and caboose of a train.

Appendix C: The Genetic Code

∞ Learning how DNA directs the cells' production of proteins is one of the great intellectual achievements of our age.

Think of the 20 amino acids that form proteins as individual words, and the completed protein molecule as a complete sentence. The arrangement of nucleotides in DNA can be thought of as a sort of Morse code, a series of dots and dashes, that can spell out the individual words (amino acids) which in turn link together to form the complete sentence (a protein).

How does the code work? It is clear that one base does not code for one amino acid, since there are 20 different amino acids and only four bases in the nucleotides. Could the code consist of two bases per amino acid? No. If you take the four different bases (A, C, G, and T) in groups of two, you get only sixteen possible combinations (4 x 4): AA, AG, AC, AT, GA, GG, GC, GT, CA, CG, CC, CT, TA, TG, TC, and TT. That could produce sixteen amino acids but not twenty.

How about a code consisting of three bases per unit? Taking the four different bases three at a time yields sixty-four possible combinations (4 x 4 x 4 = 64), more than enough to generate twenty different amino acids.

The genetic code, then, is made up of *nucleotide triplets*. Each triplet is equivalent to a set of dots and dashes of Morse code. A nucleotide triplet—or *codon*, as they are called—specifies a particular amino acid, just as the dots and dashes of Morse code can spell out a word. The sequence of codons along the DNA strand directs the order in which the amino acids are put together to form a protein, just as the sequence of words strung together forms a sentence.

All nice and neat. But nature is never quite that simple. The triplet code can yield sixty-four different combinations, not merely the twenty necessary to specify the twenty amino acids. There is a good deal of redundancy in the system; apparently, nature discovered that it is important to have backup systems a few billion years before NASA began building spacecraft.

The genetic code is simple and elegant. The four bases of DNA work in teams of three to code for the twenty different amino acids, with enough redundancy in the system to keep it working for more than 3 billion years.

Table 2 (opposite) lists the twenty amino acids, with the three-letter abbreviations used by researchers. Table 3 (page 262) shows how the triplet code of DNA produces all of the amino acids, plus special codons that tell the mRNA where to start transcribing and where to stop.

Remember, only a portion of the complete DNA molecule is transcribed to produce a protein. The same DNA molecule contains the genetic blueprint for thousands of different proteins along its long spiraling length of millions of nucleotides, so it is vitally important to be able to tell the mRNA at what point it should begin transcribing for a particular protein and where it should end the transcription. The "start" and "stop" codons, then, serve as punctuation markers.

Table 2
The Twenty Amino Acids

AMINO ACID	THREE-LETTER SYMBOL
alanine	ala
arginine	arg
asparagine	asn
aspartic acid	asp
cysteine	cys
glutamic acid	glu
glutamine	gln
glycine	gly
histidine	his
isoleucine	ile
leucine	leu
lysine	lys
methionine	met
phenylalanine	phe
proline	pro
serine	ser
threonine	thr
tryptophan	trp
tyrosine	tyr
valine	val

Table 3
The Genetic Code

FIRST BASE	U	C	SECOND BASE A	G	THIRD BASE
U	phe	ser	tyr	cys	U
	phe	ser	tyr	cys	C
	leu	ser	**T**	**T**	A
	leu	ser	**T**	trp	G
C	leu	pro	his	arg	U
	leu	pro	his	arg	C
	leu	pro	gin	arg	A
	leu	pro	gin	arg	G
A	ile	thr	asn	ser	U
	ile	thr	asn	ser	C
	ile	thr	lys	arg	A
	S	thr	lys	arg	G
G	val	ala	asp	gly	U
	val	ala	asp	gly	C
	val	ala	glu	gly	A
	val	ala	glu	gly	G

S = start codon; **T** = terminator codon

References

Ameisen, Jean Claude. "The Origin of Programmed Cell Death," *Science*, May 31, 1996, p. 1,278.

Austad, Steven N.. *Why We Age*. New York: John Wiley & Sons, 1997.

Barinaga, Marcia. "Death Gives Birth to the Nervous System. But How?" *Science*, February 5, 1993, p. 762.

——. "Life-Death Balance Within the Cell," *Science*, November 1, 1996, p. 724.

Beardsley, Tim. "Steps to Recovery," *Scientific American*, January 1997, p. 26.

Blaese, R. Michael. "Gene Therapy for Cancer and AIDS," *Scientific American*, June 1997, p. 111.

Blasco, Maria A., Han-Woong Lee, M. Prakish Hande, Enrique Samper, Peter M. Lansdorp, Ronald A. DePinto, and Carol W. Greider. "Telomere Shortening and Tumor Formation by Mouse Cells Lacking Telomerase RNA," *Cell*, October 3, 1997, p. 25.

Bodnar, Andrea G., Michelle Oulette, Maria Folkis, Shawn E. Holt, Choy-Pik, Chiu, Gregg B. Morin, Calvin B. Harley, Jerry W. Shay, Serge Lichtsteiner and Woodring E. Wright, "Extension of Life-Span by Introduction of Telomerase into Normal Human Cells," *Science*, January 16, 1998, p. 349.

Brockes, Jeremy P.. "Amphibian Limb Regeneration: Rebuilding a Complex Structure," *Science*, April 4, 1997, p. 81.

Butler, Robert N., and Jacob A. Brody. *Delaying the Onset of Late-Life Dysfunction*. New York: Springer, 1995.

Caplan, Arthur. "An Improved Future?" *Scientific American*, September 1995, p. 142.

Cerami, Anthony, Helen Vlassara, and Michael Brownlee. "Glucose and Aging," *Scientific American*, May 1987, p. 90.

Clark, William R. *Sex and the Origins of Death*. New York/Oxford: Oxford University Press, 1996.

Culotta, Elizabeth, and Daniel E. Koshland, Jr. "Molecule of the Year: p53 Sweeps Through Cancer Research," *Science*, December 24, 1993, p. 1958.

Drexler, K. Eric. *Engines of Creation: The Coming Era of Nanotechnology*. New York: Anchor Press/Doubleday, 1986.

———. *Nanosystems*. New York: John Wiley & Sons, 1992.

Duke, Richard C., David M. Ojcius, and John Ding-E Young. "Cell Suicide in Health and Disease," *Scientific American*, December 1966, p. 80.

Epstein, A. H., and S. D. Senturia. "Macro Power from Micro Machinery," *Science*, May 23, 1997, p. 1,211.

Ettinger, Robert C. W. *The Prospect of Immortality*. New York: Doubleday, 1964.

Ezzell, C. "Two Human Chromosomes Entirely Mapped," *Science News*, October 3, 1992, p. 212.

Felgner, Philip L. "Nonviral Strategies for Gene Therapy," *Scientific American*, June 1997, p. 102.

Fossel, Michael. *Reversing Human Aging*. New York: William Morrow, 1996.

Friedman, Theodore. "Overcoming the Obstacles to Gene Therapy," *Scientific American*, June 1997, p. 96.

Gosden, Roger. *Cheating Time: Science, Sex, and Aging*. New York: W. H. Freeman, 1996.

Gould, Stephen Jay. *Wonderful Life*. New York: W. W. Norton, 1989.

Hayflick, Leonard. *How and Why We Age*. New York: Ballantine Books, 1994.

———. "Myths of Aging," *Scientific American*, January 1997, p. 110.

Ho, Dora Y., and Robert M. Sapolsky. "Gene Therapy for the Nervous System," *Scientific American*, June 1997, p. 116.

Hoey, Timothy. "A New Player in Cell Death," *Science*, November 28, 1997, p. 1578.

Kinsella, Kevin G. "Changes in Life Expectancy, 1900–1990," *American Journal of Clinical Nutrition*, 1992 55: 1186S.

Kitcher, Philip. *The Lives to Come*. New York: Simon & Schuster, 1996.

Klein, George. "Malign Evolution," *Discover*, August 1997, p. 46.

Kuro-o, Makoto, Yutaka Matsumura, Hiroki Aizawa, Hiroshi Kawaguchi, Tatsuo Suga, Toshiro Utsugi, Yoshio Ohyama, Masahiko Kurabayashi, Tadashi Kaname, Eisuke Kume, Hitoshi Iwasaki, Akihiro Iida, Takako Shiraki-Iida, Satoshi Nishikawa, Ryozo Nagai, and Yoichi Nabeshima. "Mutation of the Mouse *Klotho* Gene Leads to a Syndrome Resembling Ageing," *Nature*, November 6, 1997, p. 45.

Larson, Janet E., Susan L. Morrow, Leo Happel, John F. Sharp, and J. Craig Cohen. "Reversal of Cystic Fibrosis Phenotype in Mice by Gene Therapy in Utero," *The Lancet*, March 1, 1997, p. 292.

Lee, Thomas F. *The Human Genome Project*. New York: Plenum Press, 1991.

Lingner, Joachim, Timothy R. Hughes, Andrei Shevchenko, Matthias Mann, Victoria Lundblad, and Thomas R. Cech. "Reverse Transcriptase Motifs in the

Catalytic Subunit of Telomerase," *Science*, April 25, 1997, p. 561.

Margulis, Lynn, and Dorian Sagan. *What Is Life?* New York: Simon & Schuster, 1995.

Marshal, Eliot. "Panel Approves Gene Trial for 'Normals,' " *Science*, March 14, 1997, p. 1,561.

Martin, Paul. "Wound Healing—Aiming for Perfect Skin Regeneration," *Science*, April 4, 1997, p. 75.

McKay, Ronald. "Stem Cells in the Central Nervous System," *Science*, April 4, 1997, p. 66.

Meadows, D. H., D. L. Meadows, J. Randers, and W. W. Behrens. *The Limits to Growth*. New York: Universe Books, 1972.

Medina, John J. *The Clock of Ages*. Cambridge: Cambridge University Press, 1996.

Michalopoulos, George K., and Marie C. DeFrancis. "Liver Regeneration," *Science*, April 4, 1997, p. 60.

Michnovicz, Jon J., and Diane S. Klein. *How to Reduce Your Risk of Breast Cancer*. New York: Warner Books, 1994.

Miller, Robert V., "Bacterial Gene Swapping in Nature," *Scientific American*, January 1998, p. 67.

Mirsky, Steve, and John Rennie. "What Cloning Means for Gene Therapy," *Scientific American*, June 1997, p. 122.

Pennisi, Elizabeth. "Worm Genes Imply a Master Clock," *Science*, May 17, 1996, p. 949.

———. "Chemical Shackles for Genes?" *Science*, August 2, 1996, p. 574.

Prockop, Darwin J. "Marrow Stromal Cells as Stem Cells for Nonhematopoietic Tissues," *Science*, April 4, 1997, p. 71.

Regis, Ed. *Nano*. Boston: Little, Brown, 1995.

Rose, Kenneth Jon. *The Body in Time*. New York: John Wiley & Sons, 1988.

Sterling, Bruce. "Bitter Resistance," *The Magazine of Fantasy and Science Fiction*, February 1995, p. 88.

Sternberg, S. "Course of AIDS Foretold by T Cells," *Science News*, January 18, 1997, p. 36.

Stocum, David L. "New Tissues from Old," *Science*, April 4, 1997, p. 15.

Szuromi, Phil, ed. "Genome Mapping and the Y Chromosome," *Science*, October 2, 1992, p. 9.

Travis, John. "Diabetes Results from Suicidal Cells," *Science News*, February 1, 1997, p. 72.

————. "The Science Behind the Controversial Cloning of Dolly," *Science News*, April 5, 1997, p. 214.

————. "Cystic Fibrosis Controversy," *Science News*, May 10, 1997, p. 292.

————. "Guardians of the Genome?" *Science News*, June 21, 1997, p. 386.

Wallace, Douglas C. "Mitochondrial DNA in Aging and Disease," *Scientific American*, August 1997, p. 40.

White, Kristin, Megan E. Grether, John M. Abrams, Lynn Young, Kim Farrell, and Norman Stelleril. "Genetic Control of Programmed Cell Death in Drosophila," *Science*, April 29, 1994, p. 677.

Williams, Nigel. "Thyroid Disease: A Case of Cell Suicide?" *Science*, February 14, 1997, p. 926.

Wills, Christopher. *Exons, Introns, and Talking Genes.* New York: Basic Books, 1991.

Wyke, Alexandra. *21st-Century Miracle Medicine.* New York and London: Plenum Press, 1997.

Index

Page numbers in italics indicate sidebars and tables.

272

evolution, Darwin's theory of, 122
eye lens, 65

F

Faraday, Michael, 205
Fas, 69
Fas ligand, 69
Fauza, Dario, 159
Federal Oil Conservation
    Board, 226
fetal development, 60–61
    programmed cellular death
        and, 63
    terms, *163*
fetal tissue
    cloning, 172
    research using, 171, 172,
        193, 207
    transplants, 170–71
fetus, *163*
Feynman, Richard P., 176–77
fibroblast cells, reproduction
    study concerning, 47–48
fissioning, 38–40, 45
Fleming, Alexander, 205
food restrictions, aging and,
    138–41
food supplements, aging and,
    141–42
Fossel, Michael, 66, 76
Fost, Norman, 220
free radicals, 22–26, *24*
    degenerative ailments
        caused by, 23
    natural defenses against,
        25–26

oxygen-bearing, 23
systems of body susceptible
    to damage from, 23
future scenarios
    control of research by elite
        group, 237–38
    immortality for all, 239–41
    suppressing research,
        233–36

G

Galileo, 204
gametes, 90
Gamow, George, 79
gas turbines, miniaturized,
    181
Gearheart, John D., 172
gene pool, 103
gene therapy, 17–18, 82–84,
    85, 106, 134–35
    aging and, 103–5
    AIDS and, 102
    bone marrow transplants,
        100–2
    cancer and, 99–102
    cardiovascular disease and,
        93
    cystic fibrosis and, 87–90
    diabetes and, 85–86
    Down syndrome and,
        90–92
    ex vivo method of, 83
    in vivo method of, 83
    gene pool and, 103
    sickle-cell anemia and,
        86–87

I

N

T

U

V